T0219070

Militarizing the Environment

Militarizing the Environment

Climate Change and the Security State

Robert P. Marzec

University of Minnesota Press

Minneapolis

London

The University of Minnesota Press gratefully acknowledges support for the publication of this book from the College of Liberal Arts at Purdue University.

Published by the University of Minnesota Press
111 Third Avenue South, Suite 290
Minneapolis, MN 55401–2520
http://www.upress.umn.edu

Library of Congress Cataloging-in-Publication Data
Marzec, Robert P.
    Militarizing the environment : climate change and the security state / Robert P. Marzec.
    Includes bibliographical references and index.
    ISBN 978-0-8166-9722-9 (hc)
    ISBN 978-0-8166-9723-6 (pb)
    1. Armed Forces—Environmental aspects. 2. National security—Environmental aspects. 3. Militarism. 4. Global environmental change. 5. Environmental degradation. 6. Nature—Effect of human beings on. I. Title.
    TD195.A75M375 2015
    363.738'74—dc23   2015019109

Printed in the United States of America on acid-free paper

The University of Minnesota is an equal-opportunity educator and employer.

21 20 19 18 17 16 15   10 9 8 7 6 5 4 3 2 1

Contents

Acknowledgments

Many friends and colleagues were a great source of help in the long evolution of this book. John Duvall, my colleague and coeditor of *Modern Fiction Studies* (*MFS*), was a constant support from the beginning, when I first began to think about connecting environmental issues and the security state. He encouraged me to organize a special issue of *MFS*, "The Futures of Ecocriticism," the editorial process of which proved to be a great help in expanding my original idea for the book. That project brought me in contact with a number of intellectuals whose work has proved indispensable to my own over the course of the past five years, among them Allison Carruth, Elizabeth DeLoughrey, and Rob Nixon. Rob Nixon read portions of this manuscript in its early stages and offered a number of excellent criticisms. Allison Carruth's insights on material in chapter 1, a shorter version of which appeared in our coedited volume "Visualizating the Environment" for *Public Culture*, were most informative and helpful. I owe a great debt to Mike Hill, who invited me twice to the University of Albany's Institute for Critical Climate Change and to several American Studies Association conferences to present my research on environmentality. These venues allowed me to present my ideas in public for the first time, and his intense interest in relations of ecology and war were always a great source of intellectual nourishment. Portions of research for this manuscript were made possible by generous grants from both the National Endowment for the Humanities and Purdue University's Center for Humanistic Studies.

I am extremely grateful to my editor Pieter Martin, who believed in this project from the beginning. His instructive comments and many suggestions for revision were instrumental in making this a far better book. Kristian Tvedten at the University of Minnesota Press was a great help when it came time to get the manuscript in shape and was considerably skilled in tracking down images that were difficult to find. I express my deep thanks to the two anonymous outside readers of my manuscript;

their extensive, critical, and generous comments improved the scholarship of this book immensely. So many heartfelt thanks as well to Christie Shee for her work during the final stages of production of the book. Her keen eye on the page proofs and the hours she spent producing the index were an enormous help. Finally, I thank my dear friend Bill Spanos, whose enthusiasm and personal support of my project cannot be expressed in words. His commitment to critical inquiry and intellectual standards, as always, continues to guide me—in everything I write.

Climate Change War Games

Nature is a powerful security system.

—Rafe Sagarin, *Learning from the Octopus: How Secrets from Nature Can Help Us Fight Terrorist Attacks, Natural Disasters, and Disease*

Climate change will provide the conditions that will extend the war on terror.

—Admiral T. Joseph Lopez, former commander-in-chief, U.S. Naval Forces

On July 27, 2008, the Center for a New American Security (CNAS), with the help of the United States military, members and institutes of the scientific community, private corporations, public policy institutes, national funding agencies, and ABC News, engaged in a two-day, new type of military exercise: the Climate Change War Game. The goal of the game was clear: "to explore the national security consequences of climate change."[1] The CNAS is perhaps the first in what will no doubt be a growing number of post–Cold War, post–homeland security institutions. A public policy think tank created in 2007, the CNAS's mission is to explore future threats to national security that come in the form of "irregular warfare." The original concept for the center arose out of the idea that security issues in the twenty-first century will need to be revolutionized (referred to in Pentagon circles as "the revolution in military affairs") to address twenty-first-century problems such as diminishing natural planetary resources. Its existence is indicative, as we will see throughout the course of this book, of a general trend to decentralize and expand the security society by replacing the nation-state collective fantasy of national security with the new planetary ecological-state collective fantasy of natural security. Its co-opting of environmental care and its reduction of complex ecological relations within the military parameters of a war game exemplify the primary concern of this book: the extension of the military and the national security state into the arena of environmentalism.

1

Various potentials for developing new forms of sustainability were floated throughout the early stages of the game, but as it progressed, players found such approaches to be a worrisome distraction from the central issue of security: "A focus on cutting greenhouse gas emissions runs the risk of crowding out full consideration of adaptation challenges."[2] Climate change was seen to be unavoidable and *adaptation* to the threat a more reasonable response than the exploration of new forms of sustainability. "Adaptation"—where ecosystems meet the war machine—became the command imperative to be disseminated down all lines of communication. As the inevitable end point of the war game's linear narrative, the idea navigated game players in the direction of transforming climate change into a "threat multiplier," one that trumped alternative ecological relations and possibilities. The concept of adaptation is a critical and necessary component of scientific efforts to address climate change, and it is important to recognize that salient fact. Due to considerable anthropocentric influence, climate change has already become inevitable, and adjusting to this change will be an essential historical task of humanity in the coming century. However, it is of equal importance to recognize that the concept is now functioning at a moment when humans have taken on the idea of security as a fundamental mandate of political existence. As part of a new ecosecurity imaginary, "adaptation" acquires a very different ontological character—commandeering consciousness to adopt a "clear" intelligibility that supplants the hard work of thinking alternative futures to a neoliberal paradigm. Arising out of a post–Cold War, end-of-history rationality, it tacitly obviates such diverse traditions as the wilderness sublime of Wordsworth and Thoreau, the ecstatic environmentalism of John Muir, the romantic primitivism of Jean-Jacques Rousseau, the land ethic of Aldo Leopold, the embittered inhumanism of Robinson Jeffers, and the scientific environmentalism of Rachel Carson and Barry Commoner—to name only a few.[3]

Sanctioned as a pragmatic solution to the inevitability of planetary climate change, adaptation provides perfect camouflage for military institutions bent on governing environmental anxieties from the standpoint of national security. In the age of climate change, the environment is both a source of concern (found in various struggles to implement new forms of environmental care and sustainability) and, understandably, an object of growing fear (in terms of diminishing resources, rising sea levels, growing food and water scarcity, etc.). These concerns, bound up with post-9/11 fears of terrorism, are in the process of being recast according to the parameters of risk and security—generating an environmental politics that extends the restrictive measures of homeland security into the domain of ecological security.

In addition to its pragmatic rationale, this extension of security receives endorsement at the scientific level—in enterprises such as the CIA-sponsored academic field of "natural security," a stratagem that sits ecologists, biologists, psychologists, and anthropologists down at the same table with "security analysts," "biowarfare experts," and "former spies" to develop the potential of instinctive forms of security deemed instrumental to national defense.[4] The brainchild of biologist Rafe Sagarin and security specialist Terence Taylor (former chief weapons of mass destruction [WMDs] inspector in Iraq under the Bush administration and the man who promoted the idea of "mobile" WMDs when none could be found), natural security seeks to "exploit . . . the greatest ongoing experiment in security of all time—the millions of successful defensive and offensive security strategies that abound in nature."[5] First introduced by Sagarin in the *Foreign Affairs* essay "Adapt or Die: What Charles Darwin Can Teach Tom Ridge about Homeland Security" (2003), the idea that nature holds the keys to national security has since been promoted at workshops, think tanks, and conferences in the United States, expanding internationally with the 2010 joint U.S.–U.K. military conference "Operational Adaptation: The Science and Practice of Defeating Constantly Changing Threats." There, an international community of scientists and security experts were urged to adopt adaptation as "the most important characteristic of today's international security environment."[6]

The original, ecological concept of adaptation—historically important to disciplines such as evolutionary biology (Darwin and Lamarck) and natural theology (William Paley and Leibniz)—has acquired "shock and awe" overtones in its adoption by security specialists. Coupled with its increasing appearance in ecological organizations such as the National Council for Science and the Environment (NSCE) and in journals such as *Nature* and *Science*, the concept has begun to harden into an organizational idea designed to influence political policy. According to Gary Hart, a former U.S. senator who served as chair of the American Security Project and cochair of the U.S. Commission on National Security/21st Century, the adaptive techniques of natural security reveal a form of "organic thinking" through which security specialists can discover the biological and ecological principles of defense that will enable them to tap into the militarized basis of all planetary life. Argues Hart, "It is worth contemplating the implications of an 'organic' military, one that senses in its very being, especially at the combat level, how the threat shifts and changes; what new, often crude adjustments the opponent makes; . . . how the combat 'fish' navigate the waters of their society."[7] In its new manifestation as the centerpiece of the security society's search for an organic form of defense, the environment as an inherently adaptive but equally untrustworthy being emerges as a modern codex for solving the problems of national and planetary security: How do organisms develop a defensive awareness and adapt to threats in their environment? How do dolphins protect themselves against the threat of terrorist shark attacks? How do populations of insects unify their efforts to map and secure their territories? Or, to use Hart's words again, how would "Nature behave if she were in charge of the Pentagon and our national security?"[8]

These "adaptive" maneuvers are not confined to a single security war game, nor are they an anomaly of a recent academic invention. Militarizing the environment, as we will see, is an issue that extends to all branches of the security society, appearing in multiple Pentagon documents that equally invoke adaptation as a "necessary" course of action in the "exceptional" war on global warming. The ecosystem, in other words, has now entered the national and global security imaginary (*reentered*, actually, as we'll see in subsequent chapters), and its protection and development—its governance—has become a central military concern. Natural security and the rallying cry of adaptation are symptoms of a generalized phenomenon—what I refer to in this book as *environmentality*. Environmentality is, in part, the name for a militarized mentality, one that commandeers a consciousness to wholly rethink and replace a rich, complex, multinarrative environmental history with a single ecosecurity imaginary for the post–Cold War, post-9/11 occasion. It is a pattern of thought that seeks to justify increases in national and civilian security by generating increased amounts of insecurity. As I will argue in detail in the following pages, environmentality is essentially environmentalism turned into a policing action. It names a hyperbolic mode of intelligibility, one that can be characterized as the point at which security and insecurity collapse into one another and become indistinguishable. This determining and perverse architecture is not immediately noticeable, for the self-explanatory attractiveness of adaptation and similar realistic solutions to present-day urgencies do not directly lend themselves to thoughtful historical critique, especially when it comes to addressing the immensity and complexity of global warming—an event that has its origins in more than just the Industrial Revolution.

National Security and Climate Change

I first began to notice the phenomenon of environmentality in the fall of 2007, in what I experienced at the time as a peculiar coupling of the War on Terror with a green initiative. Roughly ten months before the CNAS's televised war game, I attended a talk given at my university by R. James Woolsey, former director of the CIA. Woolsey's talk was in fact part of a tour that included, in addition to American universities, policy centers and science institutions across the United States. As well as his work for the CIA, Woolsey has had a prominent career as a national security policy leader for Republican and Democratic administrations since the 1970s. Woolsey spoke, to great applause, about the need for conservatives, like himself, to take seriously the warnings given by liberals about climate change and the need for the nation to address environmental problems in general. The central claim of his talk was that liberals and conservatives in the United States could join forces by reconstituting environmental problems under the umbrella of national security. This was the first time I had heard a conservative, who continued to foreground terrorism as his primary concern, openly argue that his colleagues needed to think about developing a green initiative. I soon discovered that he was not the only one.

Woolsey's green awakening, I realized, was a reflection of a growing and variously manifested ontological anxiety concerning the future of the environment. It differed in its specific appearance from the environmentality of the CNAS war game and Sagarin's natural security, but its commitment to militarize climate change matched the tenor of the latter and other green security initiatives. Along with a group of greened neocons, Woolsey was beginning to reconstitute the environment on the basis of its problematization as a potential target of terrorism.[9] This phantasmatic connection enabled the conservative camp to join forces with so-called bleeding-heart liberals.

Today, "going green" is good conservative politics because it means we can unchain ourselves from oil cartels and the disruption of oil-rich regions by terrorist groups and their sympathizers. Organizations such as the Green Patriots and the Set America Free Coalition present themselves as environmentalists by foregrounding their interest in hybrid, flex-fuel vehicles, arguing that this transition will free the American public from its dependence on foreign oil.[10] Prior to this, Woolsey had said little to suggest that he was concerned with environmental matters. Nonetheless, he had made it publicly clear back in 1993, after his nomination for directorship of the CIA, that the Pentagon would need to expand conceptions of security now that the Cold War was over. The United States—he argued in a mixed metaphor that sounded not unlike a B movie—had "slayed the dragon" of communism but now faced a "jungle filled with a bewildering variety of poisonous snakes" that posed even more of a security threat than before.[11]

Woolsey's use of an impression from the natural world, a flimsy but resilient, fear-producing image dredged out of colonial depictions of native land, clearly indicated that, at least in 1993, his rationale for expanding the war machine had yet to stumble upon the environment as the next monstrous threat to security. Nor, apparently, was the time right for the world's most powerful nation to consider such a minor concern, for it was concentrating its efforts on developing and installing the new threat of "rogue states." Even so, an early important venture was attempted, during the Clinton administration, to make global warming a military concern. After the fall of the Berlin Wall, developments in the military industrial complex began to shift out of their strict Cold War formations. In 1992, the CIA began to establish relations with climate scientists in a program known as MEDEA (Measurements of Earth Data for Environmental Analysis) when it declassified satellite imagery for "patriotic" climate scientists. MEDEA was the invention of Al Gore, who wrote to Robert Gates, then director of the CIA, prompting Gates to invite a group of scientists to gain security clearance to CIA documents and take a look at what the military had to offer.[12] At this point, the Department of Defense (DOD) presumably held little interest in global warming (even though the CIA had known about climate change as early as 1985 and had been working with various branches of the military to manipulate environments on a global scale as early as 1950). But Gore's commitment to addressing global warming began to generate interest. However, the second Bush administration, 9/11, the War on Iraq, and the unending War on Terror would put these concerns, for the most part, on a back burner.

Toward the end of George W. Bush's second term, however, interest in global warming exploded. In March 2007, the Strategic Studies Institute of the U.S. Army War College (SSI) held a conference on the topic—"The National Security Implications of Global Climate Change"—and produced a "Colloquium Brief" that argued the need for a "coordinated strategic communication plan" to inform and explicitly direct "public awareness." In 2008, the SSI published an edited volume of more than four hundred pages titled "Global Climate Change: National Security Implications."[13] By 2009, the institute had adopted the position that climate change is indisputable and the result of human activities. In July 2007, the Center for Naval Analysis (CNA)—the Navy and the Marine Corps's federally funded research center, which provides information for all U.S. military organizations and the government—released its first major publication directly focused on the environment: *National Security and the Threat of Climate Change*. The report maintains that the environment is a threat to planetary security and argues that the U.S. war machine must expand its power globally to avoid significant disruptions to international stability. It pushes for the construction of new military bases and command centers, such as the "African Command" previously proposed by the Department of Defense.

The impact of climate change in sub-Saharan Africa—already experiencing the effects of global warming—is of particular importance to the U.S. military and indicates the global reach of environmentality. The Department of the Army, in conjunction with the Department of Defense, began producing a series of reports in 2009 that articulated the need to establish new "Sino-American military-to-military cooperation" in the sub-Saharan region.[14] "Taking Up the Security Challenge of Climate Change" was one of the first documents produced. The document is of special interest because of its strategic creation of an environmental historical narrative that seeks to justify the ecological expansion of the war machine. In its opening declarations (a section titled "The Science of Global Warming"), the document cites the Intergovernmental Panel on Climate Change's (IPCC's) Fourth Assessment Report: *Climate Change 2007*. A United Nations organization, the IPCC is the leading international authority on planetary climate change. After the publication of the IPCC's report in 2007, the DOD began to openly characterize climate change as a threat multiplier. The report's conclusions were embraced without question, and the DOD embarked on a new offensive: deploying

Climate Change 2007 as an expedient for putting an end to the "debate" about climate change in American politics:

> Even as recently as 2006, the year in which the Academy Award–winning film *An Inconvenient Truth* . . . was released, climate change as a consequence of manmade global warming was hotly debated and deeply politicized in the United States and elsewhere. The following year, 2007, the United Nations' Intergovernmental Panel on Climate Change released its long-awaited Fourth Assessment. The IPCC report is of signal importance because it is well-balanced and moderate. It did not quell all controversy surrounding the subject; but because of it climate change is generally accepted, scientifically speaking, to be a product of manmade global warming, even though uncertainties remain as to where, when, and how much.[15]

These and other reports leave no doubt: not only has climate change been accepted, but it has also been adopted by the U.S. military as a clear and present danger. "The idea that the environment has security implications is not new. . . . What is new is that climate change poses security threats unmatched among environmental phenomena."[16] The threading of climate change—a controversial subject that was initially only articulated by environmentalists and some liberals in the United States, then only in the military by its retired generals—now appears at the most capillary levels of the war machine, reflected, for instance, in the incorporation of the phenomenon in the latest army field manuals.

These extensive preparatory representational maneuvers (and I have listed only a few here) eventually became the basis that enabled the Department of Defense to obtain a more public victory in its bid to become the nation's most dedicated environmental advocate: a congressional financial commitment for the creation of a green military. In 2010 the Department of Defense formally targeted climate change in its 2010 *Quadrennial Defense Review Report*, making a commitment to climate change key to the future of defense planning.[17] And most recently, President Obama stated in a speech to the 2015 graduating class of the Coast Guard that "climate change constitutes a serious threat to global security, an immediate risk to our national security. . . . [M]ake no mistake, it will impact how our military defends our country."[18] In this resolute radiation of environmental alarm, nature comes into existence in the narration of the security state as fundamentally a concern of the war machine, and catastrophic climate change is taken as an "engagement opportunity."[19]

Though it is beyond the scope of this present work, it is worth mentioning that this national ecosecurity framework is being mirrored in the transnational sector in such institutions as the World Bank and the United Nations. In 2012, I attended the annual NCSE conference in Washington, D.C. The NCSE has a long history of foregrounding sustainability throughout the thirteen years of its existence, but in 2012, "Environment and Security" was the title and main topic of the conference. The conference included a conspicuous number of military leaders and diplomats from both inside and outside the United States. Roughly a quarter of its main presenters were either members of the military or representatives of security institutions—including Rear Admiral Neil Morisetti (United Kingdom), Andrew Steer (World Bank), Major General (Retired) A. N. M. Muniruzzaman (president of the Bangladesh Institute of Peace and Security Studies), and Peter Wittig (German ambassador to the United Nations and a member of the UN Security Council).

At the conference, Ambassador Wittig described his efforts to introduce climate change into the agenda of the Security Council (the Security Council is understood to be *the* supreme body of the UN). His efforts, which met with a great deal of initial resistance (the Security Council traditionally concentrates its efforts on military and political security), eventually resulted in a UN presidential statement that, for the first time, affirmed that climate change was a potential threat to international peace. According to Wittig, because of his efforts and the efforts of others, climate change is now mainstreamed into the aggregate reporting system of the UN. Moreover, the UN Security Council has begun to discuss the introduction of a new global ecosecurity force: the Green Helmets. This military embrace of the environment is augmented by new economic commitments to global warming. From 2011 to 2012, the World Bank doubled its "adaptation funds" from $2.3 billion to $4.6 billion and only increased its "mitigation" (sustainability) funds by $100,000.[20] Moreover, adaptation has now become a central component in the World Bank's constitution of "development": "Adaptation is . . . not a standalone issue, but needs to be integrated throughout national, sectoral, regional, and local planning processes."[21] The International Monetary Fund and the World Trade Organization have also made similar changes to their developmental policies.[22]

These restructurings of traditional security systems and commitments of funding and resources at the transnational level suggest that securitizing mechanisms will advance in a multilateral fashion. However, nations geographically situated on the front lines of climate change, such as Bangladesh, oftentimes contain greater percentages of the world's poor. Their governments do not have the infrastructure or the civil and military might to implement adaptive mechanisms. At the NCSE conference, Major General Muniruzzaman openly characterized Bangladesh as the new "ground zero" of climate change. He argued that an anticipated sea level rise of one meter would mean a territorial loss of 20 percent, creating a refugee population of thirty-five to forty million people and a consequent destabilization of borders, resulting in a cascading effect of violence. He explained that Bangladesh would not have the resources to handle such an event and that if adaptation were to become more than an argument on paper, it would require support from nations capable of providing material investment to make it a reality. If Muniruzzaman's appeal at the conference is any indication, this expansion of the ecosecurity society will be more unilateral than multilateral in nature. And as we know, Muniruzzaman's statement concerning the climate change refugees is not the concern of only Bangladesh. It is estimated that rising sea levels will erase some forty-eight island nations and affect national borders across the Global North and South. Coupled with the loss of habitats due to hydrocarbon-influenced reductions in inhabitable land, resources, biodiversity, and so on, this legacy of neoliberalism's demand for "energy security" shows no sign of even tempering its velocity. As such, this imperial momentum promises to make ground zeros the new geography of the planet and the loss of habitation everyone's inheritance.

A Genealogy of Environmentality: The Enclosure Movement

I begin with these unsettling examples because they illustrate a recent trend in Defense Department activities, a trend that is also indicative of a momentum that extends beyond U.S. national security interests and the historical occasion of a post-9/11 America. One main concern of this book is the liminal event of anthropogenic planetary climate change—specifically the subordination of this event to military ends by the U.S. national security state and the (global) security society in general. But because the particular ecological dilemma of climate change is a *planetary* event and because ideologies of security and threats to our environment owe their birth to a much longer history, the immense scale of climate change and its coupling to Pentagon interests demands a broader kind of intellectual inquiry. Ideas such as "adaptation" and "natural security" spring from a form of intelligibility that extends back much further than these recent examples—to, most recently, twentieth-century formations of security that first became conspicuous during World War II. This contemporary period, which forms part of this book's focus, can provisionally be categorized in terms of four overlapping phases: (1) the nuclear period (roughly 1942–72), which saw the establishment, as Elizabeth DeLoughrey has shown, of the field of ecology as a discipline in the American university system[23] and which was marked by various military attempts to manipulate "enemy ecosystems"; (2) the period of "energy security" (beginning with the 1973 oil crisis), which saw the beginnings of the absorption of nature into the international stratocracy of energy production and consumption; (3) the post–Cold War period (roughly 1992–2001), which, in the reinvigoration of the "revolution in military affairs," generated more open interest in global warming; and (4) the post-9/11 period (roughly 2004 to the present), during which climate change was fully adopted as a central military concern.

But as Alfred Crosby shows in *Ecological Imperialism*, the relationship between the military and the environment spans the entire length of the modern era and extends even several centuries before Columbus's arrival in America. The post–World War II national security state's interest in the environment, its research endeavors, and the ideas it generated thus eventually led me further back in time, to the first attempt by a nation to develop a widespread system of homeland environmental security—the British enclosure movement of the seventeenth and eighteenth centuries, which, among many things, introduced modern conceptions of privatization, surveillance, and environmental manipulation. Though its ecological influence is generally recognized, the enclosure movement is not normally thought of in relation to present-day concerns for security and terrorism and the relation of these to the global spread of neoliberalism. As environmentalists have argued, the enclosure movement's national and eventual global extension of privatization effectively eliminated sustainable forms of existence in geographies across the planet—first in England; then across Europe; and eventually across parts of India, Africa, and the Americas. This privatization of land brought the environment into an imperial market economy that made possible today's forms of neoliberalism and the counterneoliberal struggles centering on the few remaining "global commons."

Frequently cited as one of the most important forms of "improvement" in land development, the enclosure movement dramatically increased England's gross national product (GNP)—well above other European nations. This movement eventually spread across Europe to France, Germany, Sweden, and elsewhere. But it was also an ecological change of the greatest magnitude, transforming land into "energy" and sustainable forms of agricultural production into new forms of mass consumption and capital development. Enclosure advocates adopted as their respective supreme goal the production of a "high yield" and the securitization of land for national and international markets. During the seventeenth and eighteenth centuries, enclosures were appropriated by the state as a national imperative, for improving nature's yields ensured the growth of the nation's population.

At this point land enters into the mechanisms of what Foucault identifies as biopower and biopolitics—that is, the regulation and control of an entire population and its territories through the profitable fostering of the life of that population. Biopower and biopolitics are terms that Foucault uses to designate a general system of human subjugation—which he calls "governmentality"—that functions not through repression but through reproductive forms that "foster life." In other words, it is the development of life as an *industry*. In his various elaborations of these concepts, Foucault does not consider ecological matters; his target is the modern development of the state and its human populations. Nevertheless, Foucault's ideas have ecocritical value (as scholars like Eric Darier, Timothy Luke, Paul Rutherford, and others have shown), and there is much in his work that can be applied equally to coeval developments taking place in the indissoluble relationship between populations and their environments. Biopower and biopolitics, I argue, can be appropriated as names for a general strategy of power that operates not through the suppression of vitalities but through the productive mechanisms of "improvement," "cultivation," and, for our ecological focus here, the "energizing" of environments—that is, the transformation of nature into a form of "energy" (to the point at which the contested idea of "Nature" is completely effaced by "energy departments" and the resultant demands for "energy security"). Unlike the more complex idea of nature, "energy" is specifically designed for a population—that is, for a conception of the human as a national and commercial interest. The idea is entirely bound to concerns that arise from the general processes of managing, developing, and expanding the politico-commercial entity of the population. The political economist Arthur Young—frequently cited as England's greatest agricultural improver—referred to enclosures as the source of all of England's political and economic power. The eighteenth-century novelist and social commentator Tobias Smollett thought that the desire to colonize foreign territories was a "strange itch" and that the uncultivated parts of England should first "be settled [i.e., enclosed] to greater advantage."[24] The national enclosures of the midmodern era enabled England to substantially increase its agricultural yields and to eventually become the world's foremost empire in the nineteenth and early twentieth centuries, transubstantiating Smollett's nationalist orientations to a transnational and multiterritorial level. The movement was indicative of a nationwide ontological shift in terms of how people related to the land and to one another. Enclosures, more to the point, constituted England's preparatory experiments in the colonization of its own land before the nation began to acquire land abroad. In the end, the transformation of a nation's entire agricultural ecosystem into commodity capital served as the cornerstone for a nation to become an empire and for an empire to transmit its power of enclosure concretely to peripheral territories. This multifaceted, isomorphic discourse of enclosure continues to function even more pervasively today, informing the West and the North's spatializations of the postcolonial world of the Global South.

However, this early international "structural development" project was more than economic in character, for it was the enclosure movement, I argue, that first coupled nature to an ideology of security. The movement's ecologically exploitative and oftentimes ruthless worldview of "development"—the demand for "land improvements" and "higher yields" that benefited the few at the expense of the many—went hand in hand with new militarized ideas and policies that ensured its continual growth. In this new discourse of enclosure, enclosed land came to symbolize the success of humanity's war against nature, made manifest in the characterization of land as an essentially unruly force needing to be brought under control. As enclosures became more of a phenomenon, parliamentary officials began to speak of them as a productive form of national defense. The extensive mapping demanded by an act of enclosure meant that the state now had a functional mechanism for gathering information about its various environments and about the people that inhabited these environments. The mapping required by an enclosure act provided the state with highly specific details about its citizens' exact locations, movements, property disputes, work habits, and general productive capacity. This transformation introduced an early form of "information processing," one that brought about a shift toward more calculative, systematizing, and security-oriented patterns of thought.

As Mary Louis Pratt has shown, the spread of systematizing practices during the eighteenth and nineteenth centuries, epitomized in the categorizing efforts of Linnaeus but also reflected in new forms of commercial enterprise, political expansion, and colonial domination, "created global [imperial] imaginings" that "operated as a rich and multifaceted mirror onto which all Europe could project itself as an expanding 'planetary process.'"[25] "The enclosure movement," she argues, "was one of the more conspicuous interventions" of this process.[26] The "systematizing of nature" reflected in Enlightenment forms of mastery such as the Linnaean classificatory tables and the massive securitizing of the land through enclosures[27] introduced a new form of rationalization that linked agricultural "improvement" to European superiority over the non-European world and urban bourgeois superiority over "nonlettered" rural communities. Enclosures and the Linnaean tables were apparatuses of mastery that introduced and regularized new methods of state securitizing, dispersing "forms of state bureaucracy that were particularly highly developed," including "record-keeping apparatuses which elaborately documented and classified individual citizens."[28] This rise of a new securitizing rationality evolved in the nineteenth century into the form of standardization that completely transformed both the military and the manufacturing industries. It was during this time that an entire constellation of developments such as these became generalized, reflecting a new *administration* of knowledge formation and production and population control. These developments at the level of thought introduced a type of state-sanctioned intelligibility that forms the basis of today's intelligence and security networks. Thus, far from being solely an agricultural event, the enclosure movement constitutes the first nationwide attempt in modernity to establish a functional and ubiquitous security society architecture on a country's population and its heterogeneous environments.

These transformations in thought, it needs to be emphasized, are indissolubly connected to current ecological issues. The embattled form of understanding involved with the rise of the security society is reflected, for instance, in later developments in the ecological sciences. The "improving," standardizing, and securitizing practices of enclosure—which reduced the heterogeneous quality of the environment to the "cash crop" imperatives of the market—are analogous to twentieth-century technological imperatives that steered scientific research toward similar utilitarian and military ends. During the early days of the Cold War, biologist and former military advisor Barry Commoner warned of the DOD's efforts to move scientific endeavors in directions that benefited the military and its industry support system. Such maneuvering, he argued, valorized a unidirectional and myopic methodology that was destructive to ecosystems. The military influence on the twentieth-century shift from classical to molecular biology, for instance, erased the focus on changing contextual relations and ecological interactions, replacing it with a mode of intelligibility that isolated and targeted discrete elements, treating them as separate from any and all ecosystemic interplay for purposes of mastering "inherent" traits (which led to the discovery of DNA, making genetic engineering and genetically modified crops a possibility). The development of this split in the discipline, he argued, led to the habitual release of chemicals into ecosystems without an understanding of how they would affect environmental interactions. The enclosing practices of the eighteenth century similarly sought to tap into nature's inherent generative capacities and supplanted diverse social and ecological relations in favor of lucrative crops and textiles. These practices would expand substantially with the rise of the Industrial Revolution. In the 1950s and 1960s, this mobilized intelligibility would ultimately lead to combined scientific and military efforts to unearth the secrets of geophysical manipulation—changing the enemy's climate, destroying a nation's crops with insects and chemicals, and using atomic weapons to melt polar ice and raise sea levels in strategic locations.

Such disparate sites of human activity are not typically connected and explored when considering events such as climate change. But as Rob Nixon has recently argued, environmental study requires a more holistic and extensive understanding of the "tentacular reach" of the political, military, and corporate bodies that impact our ecosystems: "An era of imperial overreach has brought to crisis a Washington Consensus ideology premised on the globalizing of the 'free market' through militarization, privatization, deregulation, optional corporate self-policing, the under-taxation of the super wealthy, ever-more arcane financial practices, and a widening divide separating the gated über-rich from the unhoused ultra-poor within and between nations."[29]

The events of anthropogenic planetary climate change thus challenge us to reconsider our customary efforts to understand and develop strategies for comprehending our ecological situation. As a global affair, climate change defies explanation by representative orders of interpretation. It is of a scale not previously encountered by human beings. It pushes the scholar to consider a variety of geographies, historical periods, and fields of study (and their different approaches) not normally discussed in the same endeavor. Fully fathoming the redaction of climate change to the injunctions of national security therefore demands an extensive consideration of the long ascent of environmentality as a form of intelligibility that has evolved, changed, and expanded in relation to many geographical and ecological affiliations.

Securitizing Nature: The War against Inhabitancy

This rise in a global concern for ecological security cannot be understood without first asking one additional, central question: how could humanity as a species develop, out of distinctly heterogeneous cultures and different geographies and environments, to the point of becoming a singular, homogenous, "anthropocentric" force influencing an entire planetary ecosystem? The human dream of influencing and mastering nature is, of course, not a revolutionary idea. But definite influence and actual scientific attempts to control specific environments (entire metropolitan areas, crops, the upper atmosphere, meteorological forces, ocean currents, etc.) are realizations of the modern era of military and technological history. During World War II and the first half of the Cold War, military leaders sought out scientists working in the fields of ecology, biology, agriculture, entomology, physics, and math in an attempt to extend Clausewitzian ideas of "absolute war" to the environment. Their goal was to weaponize nature on a grand scale. People like John von Neumann (a former principle member of the Manhattan Project) were highly important to the war machine. Neumann combined mathematics, game theory, physics, and quantum mechanics in various experiments to alter the climate of enemy territories. He suggested transforming the color of the polar ice caps in order to change their reflective nature (albedo) so that they would absorb rather than reflect solar radiation. Such a project, he argued, would lead to global warming and an increase in crop production in key geographical locations. Electrical and computer engineer Jay Forrester, founder of system dynamics (and who would later work with the Club of Rome in the 1970s and 1980s on issues of global sustainability), began his career by helping the military apply "command and control" concepts to large urban and social environments. He directed the Whirlwind Cold War computer project in the 1950s, which became the basis for the SAGE (Semi-Automatic Ground Environment) Air Force Air Defense System. In 1960 NATO enlisted the help of aerospace engineer Theodore von Kármán, a man who had helped design aircraft for the Austrian military in World War I. Kármán worked with chief scientific advisors of national defense organizations in the United States, Britain, Canada, France, and West Germany.[30] The "Von Kármán Committee" worked on creating "weapons of 'synoptic scale'" designed to control and dominate "whole physical systems" that "encompassed large portions of the earth."[31] The committee was influenced by previous developments in planetary control, specifically the global international data collection effort of dozens of nations known as the International Geophysical Year (IGY) of 1957–58. The mass amounts of data collected during the IGY made possible the systems of global environmental monitoring that make up today's planetary system of military surveillance. In the hands of NATO and the Von Kármán Committee, IGY information was processed by computers in order to securitize the landscape of the earth: "Had the members of the Von Kármán Committee been military historians, there is little doubt . . . they would have cast [this] as the 'decisive moment' in the history of global strategy. Time and again they called to mind the changes brought about by the advent of earth-orbiting satellites. It would prove to be, they believed, a dividing line between military eras. It promised total monitoring of the global environment, a vision of the future that was pervasive across the range of sciences and military operations."[32] Human subjectivity was expanding to become an agent of seemingly boundless surveillance and manipulation.

These developments, presumably only possible with the advent of computers and other scientific marvels, actually owed their existence to a much older mode of intelligibility that had been increasingly orchestrating human subjectivity and its environmental relations from the seventeenth to eighteenth centuries. The widespread securitizing of environments during the era of enclosures, for instance, depended on the generation of unconventional legal legislation that introduced radically new and generalized forms of species individuation and standardization—both nonhuman and human. Opposing older, more sustainable and diverse forms of human existence, these forms of individuation affected the course of human development on a global scale. As I discuss in detail, this judicial development stood at the center of the dissolution of the preenclosure legal system of *custom law* and its succession by the proenclosure and prodevelopment system of *common law*, which is the legal order that governs the human species and its environmental interactions today. The enclosure movement and common law, I argue, arose in tandem, acting like a historically determining double helix that introduced a fundamentally different understanding of what it meant to be human. This new understanding manufactured a model of the human that took as its direct enemy an environment thought to be threatening because it had yet to be properly secured, privatized, and cultivated. In order to tame this enemy, the new human had to transcend previous limitations in order to closely monitor and redirect the course of its development.

The enclosure movement, in other words, introduced a war between two "species" of the human: the commoners of the old open-field system and the individualized entrepreneurial class. This new class took as its essential burden the mastery of vast ecosystems. There was, of course, a great deal of opposition to this new human species and its enclosing policies. The rise of common law and its demand for enclosures were met with considerable resistance on the part of villagers, farmers, and laborers across England (then later by the indigenous communities in England's and Europe's colonies as the West transplanted its enclosing practices abroad during three long centuries of colonial expansion). Individuals and communities, struggling against forced evacuations brought about by acts of enclosure, invoked their time-honored rights of "inhabitancy." Inhabitancy was a political resistance term used widely until its legal status was revoked by the efforts of common-law advocates like Edward Coke (England's most powerful and historically influential judge in the seventeenth century, who almost single-handedly eradicated custom law and its tradition of inhabitancy). In other words, in addition to its restructuring of land relations, the enclosure movement also effected the constitution of human subjectivity, replacing the idea of inhabitancy with more atomized formulations of protoneoliberal human subjectivity constructed along the oppositional lines of "landlord/tenant," "landlord/laborer," "enclosed/displaced," "propertied/dispossessed," "owner/vagrant," "cultivator/cultivated," "improving landowner/unimproved land"—that is, "colonizer/colonized."

These struggles to maintain the rights of inhabitancy occurred often and consistently over the length of two centuries. And there are telling signs of these struggles today—in, for instance, the indigenous struggles against hydroelectric projects in the Amazon, the landless workers' struggle for inhabitancy rights in Brazil, the efforts of the Communal South African Land Development Project, the globalized resistant campaigns of Via Campasina (the transnational peasant farmer's organization), the opposition of the indigenous against the Grand Inga Dam proposal for the Congo River (the world's largest hydroelectric scheme), and more. These efforts constitute powerful alternatives to the ecologically destructive policies of enclosure demanded by the security society. The widespread and three-century-long erasure of the human subject as "inhabitant" (beginning with Gateward's Case in 1608), its connection to the enclosure movement, and its centrality to the development of *common law*—each of which transformed human–environmental relations into the economics of "improvement," "development," and private property that are operative today—constitute the background for the securitizing and militarizing of environments found in twentieth- and twenty-first-century technological commitments. The new species of the human these processes created acted as a singular force, compelled by individualized, economic, and biopolitical demands to transform environments into sources of security, energy, and wealth.

The importance of these three—the eradication of inhabitancy, the development of common law, and the mandate to "improve land" that defined the enclosure movement—cannot be overemphasized. They inform, for instance, the writings of such giants as Adam Smith (1723–90), John Locke (1632–1704), and Thomas Hobbes (1588–1679)—Enlightenment thinkers and political economists who paved the way for the architecture of neoliberalism. All three wrote in the midst of these enclosing transformations. And all three, despite their differences, developed notions of statecraft grounded in an assumption of "human nature" (as warring, selfish, and only interested in personal gain) that organized its struggle and freedom around the cultivation/subjugation of the earth. Locke, an advocate of the enclosure movement, developed private property in his writings as the relational mechanism of societal security, and Smith's *The Wealth of Nations*, which characterized the human subject as an isolated individual motivated to maximize his or her personal interests (*Homo economicus*), became, of course, the universal standard for modern-day capitalism. Their conceptions of human independence, moreover, also arose in the wake of earlier Enlightenment advancements in what would eventually become the formal sciences—in particular, the "new physics" developed by mathematicians and physicists such as Copernicus, Galileo, and Newton. (These developments in the formal sciences would become crucial to information gathering and security during and after World War II.) This new mathematical science of physics, which sought to emancipate human beings from their dependence on chance, brought about a form of securitization that changed the understanding of Nature from an entity on which one depended into an entity that posed a threat. The "natural world" became associated with the threatening ideas of contingency, risk, and endangerment, and "freedom" became an idea that involved the act of becoming *independent* from these conditions. From the standpoint of these formalist theories, the environment had to be subdued—securitized—for humans to obtain their independence. This new mathematical mastery of chance informed the new political economy and social contract theory of Locke, Smith, and others. This powerful intellectual combination essentially ratified the idea that humans had to emancipate themselves from their (unenclosed) environments. As such, this constellation of events helped generate a singular momentum, one that makes possible the present-day structures of environmentality.

In the nineteenth and twentieth centuries, the discourse of enclosure evolves, becoming more militarized and branching out to realign and newly channel its energies in accord with scientific and technological developments. The scientific management theories of Taylorism and the general economy of standardization embraced by the army and the factory system would soon reappear, in a slightly different form, in the emerging national enclosing mechanisms of air surveillance, mapmaking, and other information-gathering technologies of World War I and World War II. Expansions of organizational methods of standardization and mechanization begin to couple the isomorphic discourse of enclosure with new forms of disciplinarity, militarization, and securitization (chapter 3). These developments reach a kind of apex in the twentieth century, with the rise of the military industrial complex in the United States (chapters 3 and 4). At this time, science (certain disciplines like biology, engineering, and physics), the military machine, and the nation-state begin to work in tandem as parts of the same organizational motor, impacting environments unlike ever before. These affiliations, as I will show, are operative some two decades before anything like a general environmental awareness, study, and movement come into existence in the United States. By the time of the publication of Rachel Carson's *Silent Spring* in 1962, Barry Commoner's *Science and Survival* in 1967, the 1960s rebirth of the influence of Aldo Leopold, and the first Earth Day in 1970, this triumvirate of the sciences, the military, and the nation-state had already been researching and establishing essential connections to the environment for decades.

Eventually, as the United States devotes its attention to Cold War issues of nuclear development, the space race, and the general post–World War II expansion of the security society, it begins to take a direct interest in the environment. Long before Sagarin began his search for the security blueprint of life in ecosystems, the military had been researching the study of specific environmental life forms in the hopes that they would find in nature organic disciplinary and security-oriented organizational tendencies. The formal sciences, like mathematics, which were important to unlocking enemy codes during World War II, began to work with the natural sciences to uncover these secret codes of nature. The combined militarization of mathematics, physics, chemistry, and molecular biology offered the security society avenues for gaining mastery over the essential codes of existence. Unlocking the secrets of the atom was, of course, one of the first priorities. At the same time, the effects of toxic fallout from experiments in atomic and nuclear warfare introduced an idea that the enclosure movement had practically erased from history: that discrete elements in nature might affect and be affected by one another and be part of a larger eco*system*. Such a system, members of the Atomic Energy Commission considered, might be an appropriate field of study at the university level. Modern-day ecological "awareness" thus first develops when scientists involved in nuclear defense projects begin to confront the destructive elements inherent in military-oriented technological development. The radiation from nuclear experiments that devastated environments led to new interests in the *systemic* nature of ecosystems. In both cases—unlocking the secrets of nature and discovering nature's relational essence—the environment was constituted within a militarized horizon as an expansive and instructive "laboratory." Long before environmental organizations began to warn the public about "local" actions impacting a "global" world, the military had been impacting and scrutinizing these very environmental interactions.

This specifically imagined ecological horizon is incorporated by the state as a kind of stage, within a larger military theater, for generating new methods of securing and governing the general population. At this middle period in the historical development of environmentality, it is organized for the most part around Cold War concerns—in other words, the need to secure the homeland population and its allies against the worldwide expansion of Communism. As a laboratory, the environment continued to be constituted as it was during the great eighteenth-century age of enclosures—as an unlimited source of technological and economic wealth. But new military interests in organization and new commitments to the securitizing of the population generated efforts that began to embrace an expanded conception of the environment. The natural sciences (biology, physics) and the formal sciences (computer science, information theory, game theory) each began to take on significance as the search for new forms of control, surveillance, and governing become part of the nation-state apparatus. The developments in these fields coalesced within the horizon of a certain understanding and constitution of the environment. The environment was reterritorialized along a contradictory but reciprocal axis: as a site that yielded presumably natural forms of governmentality but also as an unstable site that needed to be logistically supervised and managed.

At this Cold War stage, questions of environmental degradation (endangered species, climate change as a threat to the future of the planet) still lie beyond the horizon of this environmentality. The environment existed as an object of investigation and manipulation, not as an object of concern. With the fall of the Berlin Wall, the military-industrial-state complex gradually reconceptualizes the environment again, this time as an issue of concern directly relating to national security (manifested first in Al Gore's attempts to charm the CIA into accepting climate change as one of its mandates). However, it is not until the combination of a public weariness over the "unending war against terror," the problem of increasingly poor relations between the United States and other nations in the wake of the preemptive war in and then occupation of Iraq, and the publication of the United Nations' Intergovernmental Panel on Climate Change's

Fourth Assessment Report (2007) that the future of the environment conspicuously enters center stage as a matter of national security. This new entrance of the environment into the military-industrial-state order constitutes the full appearance of environmentality—an ontological ground plan that had been developing gradually since the imperial expansion of enclosures.

This is not to say, however, that the brief genealogy I have laid out here is linear or continuous. There are, of course, many turnings and counterdevelopments to this environmentality, just as there were many acts of resistance and forms of resistance to colonial expansion. This highly influential discourse is not an inescapable metanarrative imposing a monolithic totality on the diversity of national formations, nor does it make impossible many other forms of environmental explorations, representations, and struggles. No order, despite its strength and durability, can be said to be absolute. As I have suggested and will explore in detail in the following pages, there were many alternatives and active resistances to this process. Nonetheless, to discard the continuity and weight of certain decisive patterns of thought on pluralist grounds would be to reject an understanding of history as a series of unequal developments that seriously impact our current state of affairs—climate change being one of these states. If long historical influences were not a part of human development, we could never speak of anything like the agricultural revolution or, more recently, the ideology of imperialism that supported the Western colonization of presumably inferior races. Environmentality thus is the result of a long series of exploitative and militarized relations to ecosystems, and its purported material domain—the environment—is a territory of human existence that has, like other fields of inquiry, been brought within the governing mechanisms of the state.

This book, therefore, rejects the idea that the "greening" of the war machine and the attendant dissemination of adaptation as global humanity's new postpolitical, end-of-history rallying cry are simply logical responses to our current historical moment. Nor does it interpret this greening as the symptom of a recent trend in military philosophy. It is reflective, I claim, of a less recognized and more latent genealogy of security that underpins America's and the West's relationships to their various environments. That trend extends back, in part, to a certain ontological imperative that has its origins in the seventeenth- and eighteenth-century securitizing of the commons—an imperative that was subsequently adopted by the United States in various forms throughout the nineteenth and twentieth centuries as it supplanted England as the reigning global power.

The history of the enclosure movement and its development as a widespread discourse is of course highly complex, sporadic, and heterogeneous in nature—operating at different scales, from the early days of Henry VIII's Dissolution of the Monasteries to, for instance, the nineteenth-century British colonization and enclosure of Nigeria. But it should be recognized that as investments from the many various registers of human production (economic, political, military, etc.) began to course through the enclosure movement, the power of its influence grew. To a large degree (to cite just one example) the previous century's colonization of Nigeria made possible the twentieth-century privatization of Nigerian oil by military Junta supporters Chevron, Mobile, and Shell (companies that further intensified the enclosure of Nigerian resources, signed security contracts with Nigerian state police, imported handguns and ammunition for police and security officials, and hired armed security forces to fly by helicopter to fire on Nigerian environmental activists demonstrating on an offshore drilling platform).[33] Despite the complexities of this history, the ideal of *national security* that resides at the heart of the discourse of enclosure functioned as a continuous and decisive imperative from the seventeenth century on, when the British government decided to oversee the movement, centralizing individual acts in the form of amalgamated Parliamentary Acts of Enclosure. In the wake of these developments, security not only became a central political and economic principle but also served as a new essential ground of thinking. As the enclosure movement gained prominence, this development in security generated a redoubtable anthropocentric will-to-power, one dedicated to an aggressive technique of developing and controlling nature and the world's environmental resources—at almost any cost.

As I have indicated, environmentality as a critical term owes its conceptual development in part to Foucault's explorations of "governmentality" and "biopower." But I also intend for it to refer to the systematic, modern-day belief that humanity has achieved a level of development beyond any and all ideological restrictions. In the discourse of environmentality, this is manifested in the supposed discovery of innate forms of development and security. These discoveries claim to have broken through all forms of political obfuscations and ideological agendas by providing direct access to the inner truth of reality. The common-sense solution of adaptation is found to be desirable not only by the state but by its citizens on the grounds that it speaks to one of the key desires of a postpolitical, end-of-history population: adapting to the impending event of sea level rise, for instance, appears in all obviousness to be the clear-headed answer that cuts through not only the muddle of the "debatable" status of global warming but also the red tape of "big government" that keeps individual subjects from realizing their own freedoms and desires. This dynamic is operative in such declarations as the following statement made by Phyllis Cuttino, director of global warming at Pew Charitable Trust:

> The military isn't waiting while Congress and the general public are having their debates about climate change. They're stepping out as they have on so many other things. . . . If there's anybody that's going to be at the forefront of how to save energy, reduce global greenhouse-gas emissions and become more efficient, it's [the military], because it's in their best interest. . . . If they can do it well, it proves to the rest of us that we can do it well. . . . We're all going to benefit from what they're doing.[34]

The actions of the military, security institutions, and certain members of the scientific community in this sense find a convincing purchase by standing in opposition to the traditional body of the government with all its bureaucracy and endless haggling of partisan interests. In this way, the authorities of the security society partially occlude their status *as* authorities and appear also and equally as independent desiring subjects speaking on behalf of other desiring subjects wanting to free themselves from the traditional bonds of the state. In this sense, environmentality enables the military and the security society in general to characterize themselves as paradoxically part of the state but also not part of the state. They are part of the state because they are political organizations. But they are also not part of the state in that they are the ultimate assemblages not of control and freedom-reducing rules and regulation but of *action*: the security state becomes the war machine that takes action while the *other* state—big government—continues to curtail the freedom of individuals by imposing, for instance, more "environmental legislation" directed at decreasing their production of greenhouse gases. In this precise manner, ideology and clear reality constitute two sides of the same coin, for the ultimate ideology is the belief that one has successfully broken through all ideologies (i.e., all curtailments of freedom and all limitations to humanity's direct access to nature). The security society, in other words, occludes its ideological constitutions by defining its activity as an organic rationality and as a self-evident practice of confronting the ecological crisis of climate change (deploying "adaptation" as Nature's way of living with its own aggressions). Environmentality therefore is also the name for a nonpolitical political movement made possible by the very logic that has produced the various forms of ecological crisis we now confront. In the face of the "end of history" denegation of politics by the logical economies of the neoliberal state and its war machine—economies that foreclose a proper political act (in Rancière's sense of the word, as an intervention that would change the very framework that predetermines how things work)—environmentality names an administration of ecological matters from the standpoint that only by remaining within the existing (neoliberal) frame of our politico-environmental relations, only by providing "practical," "real world," "organic" solutions (namely, only by accepting the "inevitable truth" of adaptation) can we hope to address the problem of climate change.

This position, as I will explore, is not as clear-cut as the security society would like us to think. On the one hand, there is certainly much to be said for a no-nonsense acceptance of climate change by the military as a new and genuine obligation: it may help to instill a general awareness of climate change and the threat that global warming poses to our planet's growing ecological fragility; it may bring about a new widespread sense of environmental responsibility, and it may break through the remaining and still decisive conservative view backed by corporate petrochemical interests that climate change is a hoax. However, its disavowal of its ideological composition seriously disadvantages the potential for alternatives and can result in the reduction of the state to an aggressive ecological policing agent. Its complicity with neoliberalism can lead to a new and even more vicious reinforcement of market forces over the needs of the planet's dispossessed and their environments. Its specific form of atomized scientific objectivity can make the solutions, say, of genetic engineering efface alternative philosophies that seek to encourage the reconstitution of human subjectivity toward more environmentally responsible formations. And, ultimately, it may, in the face of a struggle over diminishing resources, impose the defense-oriented solution that seeks to remap the earth along the lines of a gated community—thus transforming entire bioregions into enclosed, segregated zones or, to speak like Agamben when he was discussing the policies of the Bush administration, into the camps that will arise when the "state of exception becomes the rule."[35] If the history of the twentieth century—the century of the concentration camp—is of any indication, the representation and reorganization of the complexities of the environment under the reified logic of security can only lead to horrifically predictable ends. Such postideological understandings, therefore, no matter how compelling, cannot go unchallenged if we are to genuinely confront the twenty-first-century age of climate change.

Despite these reservations, the purpose of my critical analysis, I want to be clear, is not to deny the relevance of patient and sober scientific work generated by organizations such as the IPCC and other concerned environmentalists and ecologists. The point instead is to rescue this urgently needed knowledge from the security society's shock-and-awe tactics, a methodology that can only result in the further advance of an enclosing instrumentality. It must be recognized as well that the geological event of global warming is not an innately occurring event of biological processes but the outcome of a certain evolution of the human species and the subsequent march of that species formation through history. This movement has led us to the point of generating the new geological age that Paul Crutzen has aptly promoted as the "anthropocene."

The IPCC has been extensively outlining the anthropocene's multifarious ecological changes for more than two decades. *Climate Change 2007* confirms that global warming is "unequivocal" and that it is primarily the result of human activities, with these activities generating the greatest output of greenhouse gases between the years 1970 and 2004 (a 70 percent increase since preindustrial times). All its predictions are grim: from increased water stress and food scarcity in locations across the planet, to the significant loss by midcentury of the eastern Amazonia forest, to sea level rise and the consequent erasure of the planet's forty-eight island nations, to the production of millions of "climate change refugees," to the alarming event of mass extinction (with as much as 30 to 50 percent of all species dying off by midcentury—what ecologists now refer to as the sixth massive planetary extinction of animal and plant species, the worst period of die-offs since the dinosaurs). Current commitments to mass energy production and consumption, habitat destruction, overfishing, topsoil loss, dam production, pesticides, and a host of other entropic, "collateral" aftereffects of anthropocentric development make this event unlikely to change in the foreseeable future. Are we to understand, in listening to the security custodians of the fantasy of humanity's biological destiny, that the truth of human existence is no more than the necessity of adapting to this spiraling ecological destruction? Surely there are other alternatives to the security imperatives of the "accept things as they are" logic of this environmentality.

It would seem, though, that the search for alternatives—the entire orientation of environmental movements in the 1970s, 1980s, and 1990s—increasingly receives less attention and even less acclamation. Instead, spectacularized threats to our environmental future constitute the predominate approach in documents produced by the world's governing legislative and corporate bodies. These deployments of insecurity ensure the continual increase of military budgets and the search for even more asphyxiating forms of security. By 2008, when green military narratives first began to appear in noticeable force, the persuasive effect of the Bush administration's War on Terror narrative had ostensibly begun to wane both politically and publicly. Yet the structures of security generated during those years continued to expand. The policies that defined the Bush years—legalizing preemptive war; promoting an unending War on Terror that devoured vast amounts of resources and wreaked havoc on the economy; regularizing racial profiling, indefinite detention, and torture; suspending civil liberties; normalizing illegal wiretapping and surveillance; and instituting drone warfare—continue to define the global political economy of the present. The Obama administration maintains its commitment to drone surveillance and warfare, devoting billions of dollars to their development. Enterprises such as the National Security Agency Dishfire program expand unabated, despite growing criticism, collecting close to two hundred million text messages a day and tracking e-mail and Internet usage on people across the planet. And the U.S. War in Afghanistan shows no clear signs of coming to an end. More than thirty thousand U.S. troops and close to twenty thousand allied forces remain, with the Pentagon pushing to maintain these numbers at least until 2016.

Exasperated by the realities of a failed Kyoto Protocol, escalating CO_2 in the atmosphere, vanishing polar ice caps, species extinction, and more, the IPCC and a host of other environmental agencies began to ramp up their efforts to change public opinion and political policy in the North and

West. Generally ignored throughout the eight-year reign of George W. Bush, the IPCC's reports—the product of five hundred authors representing more than fifty nations and more than a dozen international organizations and scrutinized by some two thousand expert reviewers—now reach a global audience. A Google search for the IPCC's latest report, *Climate Change 2013*, already produces more than 770 million results, whereas the previous report, *Climate Change 2007*, produces 147 million. Yet almost all the military and security documents that I have mentioned also directly cite the IPCC's reports and ennoble them with unquestionable justificatory emphasis. These military documents use the IPCC as an exemplary scientific endorsement in a general campaign to declare the environment the next significant threat to national and global security.

There is no question that policymakers need to genuinely confront the seriousness of the situation and the facts laid out by the IPCC and other organizations. But are these scientific findings and projections, coupled with the changing national security context of the twenty-first century, the only or even the primary kinds of formulations needed in order to change the course of the anthropocene? In one respect, such formulations are still sorely needed. After all, there are still many conservatives and moderates (and even liberals) that continue to deny the reality of climate change (as witnessed in recent Gallup polls[36] and in legislative acts, such as the U.S. House and Energy Committee's vote to reject scientific findings that support global warming as an actual fact of our existence,[37] the Inhofe-Upton Bill to block the Environmental Protection Agency,[38] and the South Dakota House Representatives Resolution that urges public schools to teach climate change not as hard science but as a theory equivalent to astrology).[39] However, an approach that just focuses on "getting out the facts" assumes that environmentalists will be able to save the environment and win the struggle for environmental justice by simply compelling conservatives—and certain moderates and liberals—to "finally face up to the realities of global warming." Such a battle, however compelling and important, only scratches the surface of our current historical occasion. It fails to uncover the driving forces—the more extensive historical and ideological structures—that have generated, defined, and maintained our theoretical and practical approaches to climate change within the fantasy of environmentality. It is to these forces that I now turn.

1

The SAGEs of the Earth and the Accidental Nature of Environmentality

Nobody believed you could contaminate the world from one spot. It was like Columbus when no one believed the world was round.

—An unnamed scientist working for the Atomic Energy Commission after discovering that nuclear testing in the Pacific had contaminated the planet[1]

The security society began officially in the United States when President Truman signed the National Security Act of 1947. The act created the Central Intelligence Agency and merged the War Department and the navy into the National Military Establishment, which eventually became the Department of Defense in 1949. These agencies, in addition to their instrumental role in the formation of Cold War strategies of information gathering and deception, extended the apparatuses and discourses of security and knowledge production that had started to define the character of the military industrial complex during and prior to World War II. But already—that is, before the official establishment of the security society— the paradox of formulating a network of comprehensive security that depended on the active constitution of life as an "ecosystem" of widespread *insecurity* had already begun to operate as an obligatory, driving force of intellectual inquiry. This essential fantasy—fabricating security by generating insecurity—is part of the nature of environmentality. As I argued in the introduction, the contemporary era of environmentality comprises four overlapping phases: (1) the period of atomic and nuclear development (roughly 1942–72), (2) the period of "energy security" (beginning with the 1973 oil crisis), (3) the post–Cold War period (roughly 1992–2001), and (4) the post-9/11 or "climate change" period (roughly 2004 to the present). The first historical period, despite its different historical context, contains the seeds for the schemes of *adaptation* and *natural security* that define the developments of green militarism in the present. The coupling of security and insecurity that defined this period first became visible during the development of the Manhattan Project—*the* enterprise that came to circumscribe and define the strategy of the war machine and its future development. The Manhattan Project solidified what President Eisenhower later referred to as the military industrial complex by tightly weaving military concerns together with industry and university science departments. These connections favored a style of intellectual inquiry that was seasoned in the art of combat—militarizing scientific and technological research from the ground up.

The Manhattan Project began officially in 1942. However, as early as 1939, the infamous "Einstein-Szilàrd letter" (to President Roosevelt) firmly planted the idea in the minds of state officials that a new form of mass destruction was "certain . . . [to] be achieved in the immediate future."[2] The need to successfully invent a weapon of mass destruction subsequently became the defining teleology of the war. In addition to its emphasis on the need for American scientists to develop the potential of this mass destruction ahead of Germany, the letter encouraged Roosevelt to expand the limited funding and resources of American universities by making it possible for universities to work with industrial laboratories. Even later, when it became clear to the U.S. War Department that Germany was nowhere near the stage of developing a working atomic bomb, the ideal of reaching total securitization (winning the war in no uncertain terms) through the technoscientific development of a weapon of immense destruction (total insecurity) continued to wholly determine warfare strategy.

This unwavering pattern of thought was not only an ideological demand enforced by the Roosevelt administration; it was also a demand shared equally by the scientists working on the Manhattan Project. In his recent study of the dynamics of American warfare from World War II to Iraq, historian John W. Dower shows how this pursuit to harness forces of extensive annihilation—forces never before attempted by a single species—depended on the quasi-religious belief in a unidirectional pattern of thought. The pursuit became a relentless obsession and generated the consuming idea in the minds of scientists such as J. Robert Oppenheimer, Victor Weisskopf, Enrico Fermi, and others that discovering the secret to mass destruction was an unstoppable "organic necessity" inherent to scientific inquiry itself. This is how Oppenheimer characterized it in 1945:

The reason we did this job is because it is an organic necessity. If you are a scientist you cannot stop such a thing. If you are a scientist you believe that it is good to find out how the world works; that it is good to find out what the realities are. . . . It is not possible to be a scientist unless you believe that the knowledge of the world, and the power which this gives, is a thing which is of intrinsic value to humanity, and that you are using it to help in the spread of knowledge, and are willing to take the consequences.[3]

In his attempt to justify the construction of the atomic bomb—an attempt that sought to unlock and harness the secrets of nature—Oppenheimer shines a light directly on the core design of thought informing today's environmentality. We should not simply take this as the intellectual temperament of a mind governed by a monomaniacal will-to-power. As Dower himself effectively reveals, Oppenheimer's statement reflects a general attitude, one that replaces the idea of "just war" with "total war."[4]

Dower's separation of "just" and "total" suggests a fundamental shift. However, as we will see, the line between just wars and total wars had already become indistinguishable. From the seventeenth century onward, justice was increasingly associated with the idea of a fullness of control, beginning with the establishment of common law and the enclosure/colonization of vast environments. In the twentieth century's production of closed systems (closed off from criticism, closed off from changing the course of an intellectual pursuit), total war generates forms of destruction that are either considered to be just because they are necessary or accepted as an inevitable part of confronting the hard facts of reality. To return to Oppenheimer's statement, the scientist true to his or her profession will assume the mandate of unveiling the secret nature of reality (performing hard science by confronting hard reality without prejudice)—as if the reality of releasing a force of mass destruction preexisted its own constitution in and through the intellectual activity of a wartime-motivated inquirer. Genuine knowledge production, according to this logic, involves the uncovering of a positivistically verifiable order of being. The empowerment of this knowledge as undeniable gives it its "intrinsic value." The totalitarian nature of such a conception of true knowledge—its status as unquestionable—is even overlooked when the preordained goal of that knowledge is the destruction of a society and its environments on a monumental scale: the inquirer on the journey to discover the ontological core of reality must be "willing to take [the] consequences." At such a point, the difference between a just and a total war collapses into the act of producing a style of "knowledge" governed by a specific demand: the victory of total security is achieved through the inauguration of massive insecurity as the basis of existence.

Oppenheimer's Ahabian viewpoint reflects a general attitude. The fidelity to mass destruction defining scientific activity toward the end of total war would expand to influence various forms of seemingly justified "total war" research and development in subsequent years. The race to develop the atomic bomb that defined the last years of World War II led to an exponential growth of nuclear testing in the 1950s and 1960s. Security (against the Soviet threat) and insecurity (the successful production of enough nuclear weapons to destroy the world multiple times over) would so collapse into one another as to become indistinguishable. In addition, the attraction of atomic energy's vast destructive power spilled out into a considerable number of biological and ecological experiments. Fearing that the Soviet Union would be expanding its weapons arsenal, the United States began to explore the potential of biological weapons, experimenting with pathogens capable of destroying large population centers more effectively than the atomic bomb. As military historian Jacob Hamblin reveals, scientists and military leaders "had agreed that in a total war, disease pathogens could be effective instruments of human death. . . . They had chosen which pathogens to focus on, . . . begun field trials, and . . . built production facilities for the ones likely to cause widespread human death most efficiently."[5] Once such pathogens were released, military leaders knew that they would be impossible to control. Nonetheless, the risk was deemed necessary. In 1949 Secretary of Defense James Forrestal hired entomologist Caryl Haskins to chair a committee devoted to exploring the potential of biological warfare. Forrestal wrote to Haskins, encouraging him to "undertake a full examination of all the technical and strategic possibilities of biological warfare."[6] The committee began to research the potentials of deploying pathogens in order to destroy enemy crops. In the late 1950s, scientists sought new ecological forms of warfare through the manipulation of the world's climate, speculating that the Soviet Union "might one day be capable of freezing Florida or turning the Midwest into a permanent drought belt."[7] And soon after being formed in 1949, NATO further expanded the destructive reach of total war by considering the potentials of cataclysmic environmental destruction, going so far as to suggest the use of nuclear bombs to reconfigure the sea floor and change the course of ocean currents.

The unquestionable adoption of an inevitable course of action, of chaining "knowledge" and "value" to an unstoppable force (the construction of the atomic bomb, the stockpiling of nuclear weapons, the destruction of entire crops, and, most recently, the proclamation of adaptation) leads me to the central and most troubling feature of environmentality: the normalization of ecological destruction—of ground-zero accidents—as an inexorable, even compulsory aspect of contemporary existence. In his analysis of the technological character of modernity, the French urban theorist and critic Paul Virilio gives special emphasis to the modern-day accident of technology's destructive aftereffects. Virilio is known well for his work on the relationship between the modern passion for speed and the effect this passion has on an increasingly self-destructive form of politics. He is also perhaps the first theorist to suggest that ecological concerns are directly caught up in the workings of the war machine (in his 1978 *Popular Defense and Ecological Struggles*). Virilio's latest theoretical work is of special interest when it comes to analyzing the forces influencing our current ecological occasion, for it can help us denaturalize the complimentary relation of total war to the fantasy of total security.

Virilio's works suggest that the more the idea of total war is fastened to the idea of an unstoppable intellectual inquiry, the greater the potential for large-scale accidents. Typically, accidents are comprehended as unwanted and in some cases unintentional side effects of modern technological development. Virilio, however, argues that the accident should not be explained away as unfortunate or coincidental; rather, it should be properly identified as a fundamental marker of twentieth- and twenty-first-century reality. In his *Unknown Quantity*, he argues that there is a direct relation between technoscientific production and the increase in the number of local and global accidents. Within the scientific world picture, which presents a certain idea of "organic" thought and technological development as the highest achievement of human creativity, the accident is represented as an unfortunate but necessary side effect—as an "error" in human judgment or design, as a limitation in the reliability of materials, or as a necessary evil. As such, the accident is comprehended as an event that can be overcome through an increase in the security of procedures (e.g., the reduction of "collateral damage" through enhancements in targeting technologies or the preservation of soldiers by their replacement with drones).

Virilio's "accidental thesis" directly opposes this view. His point is that in today's world, the accident, oddly, is *not* accidental; it is, rather, "becoming a clearly identifiable historical phenomenon."[8] Despite the increase in security measures (which has spawned its own field of intense study and development in almost all disciplines of the sciences), the recurrence of accidents has intensified. More to the point, late modernity's particular technoscientific mode of production, Virilio argues, releases the accident into existence as a primary and *required* component of reality. No longer an occasional offshoot of worldly manufacturing, the accident arises out of a fundamental change in the essence of modern human production: "The invention of the substance is also the invention of the 'accident'" (6). The material substance of technologically structured reality has internalized the accident as part of its very composition: "The shipwreck is indeed the 'futuristic' invention of the ship, the air crash the invention of the supersonic plane, and the Chernobyl meltdown, the invention of the nuclear power station" (6). Virilio, consequently, defines ecological disasters as the "clinical symptom" of the enfolding of the accident into the substance of existence (7).

This folding of the accident into the substance of existence forces us to rethink its status as defined by the discourse of technical rationality. "Substance" here indicates both a style of technology inherently connected to accidental production and a state of mind. In its constant "advancing," the substance of technology outpaces human understanding of its precise workings. Thus the curious activity of accepting products without understanding how they work becomes a naturalized state for the human user (the average consumer can use and understand the function of material instruments but cannot fix these items, since the specific knowledge of how they operate is not part of consumer consciousness). Instrumentality thus disappears from human consciousness, from the "field of the visible," to speak like Althusser. Quoting Paul Valery, Virilio forces us to consider a significant point: with the slippage of instruments into the automatism of habit—the acceptance of greater sophistication with greater incomprehension—the only aspect of the visible that enters consciousness is the *breakdown* of the instrument. It is the accident and the accident alone that fills out the field of the visible: "Consciousness now exists only for accidents" (6). In this development, in which consciousness is repositioned to focus on the accident, *existence is transformed into a recipe for producing and maintaining a constant state of anxiety.*

This in turn produces the feedback loop of a need for increased security—the problematic solution produced by a system that can never get outside of itself. We thus find ourselves in the absurd, precise opposite position of the kind of instrumental breakdown thematized by Heidegger (in his famous example of the broken hammer that suddenly unchains the subject from the normalized processes of production and reproduction). The breakdown of the instrument no longer opens us to the encounter with the ontological essence of existence—the confrontation with, as Heidegger phrases it, "the being of being"; as instrumentality disappears from consciousness, its breakdown can only confront us with the terror of the accident. To put this more emphatically, in today's world, broken devices no longer carry the power to enable us to realize the ideological underpinnings of existence. Breakdowns are not the cessation of the specific substance of existence that colonizes our lives; *breakdowns are the primary substance of existence itself*—the confirmation of and justification for more technological advancement, "states of exception," and security.

Virilio thus tells us something crucial about the status of the environment-object world that was constituted as a result of the obsession with total technological domination: "In a world which is now foreclosed, where all is explained by mathematics and psychoanalysis, the accident is what remains unexpected, truly surprising, the unknown quantity in a totally discovered planetary habitat, a habitat overexposed to everyone's gaze, from which the 'exotic' has suddenly disappeared in favour of that 'endotic' Victor Hugo called upon when he explained to us that, 'It is inside ourselves that we have to see the outside—a terrible admission of asphyxia'" (129). This aspect of an "asphyxiating overexposure" that ultimately follows in the wake of the foreclosure of a system is an aspect that needs to be emphasized when it comes to analyzing military environmental interventions. When a system is taken for granted, and the knowledge that we are in an ideologically constituted system falls into oblivion, the search for fundamentally different systems fades into oblivion as well. Hence the only action left is to increasingly overexpose and develop what already exists. This drive to "overexpose" for purposes of total (final) explanation and control, technological manipulation, and economic development (the high-yield logic of the discourse of enclosure)—the very logic that informed the Manhattan Project and that now informs the deployment of adaptation—creates a world of increasingly dangerous and globally consequential accidents.

We can extend Virilio's work here by first drawing a firm connection among the accident, the environment, and what I will develop in chapter 2 as the "war on inhabitancy." The Love Canal disaster in the United States, for instance, which galvanized a specific environmental awareness and movement, is one example of an "accident" that irreparably damaged an environment and its inhabitants. But the accident is more firmly connected, even embedded now, in the planet's ecosystems in the late enclosing age of postmodernity, as we can see in the profit-oriented overproduction and subsequent destruction of environments valued for their resources. The invention of the oilrig is the invention of the oil spill. The invention of the cash crop for the neoliberal, global market is the invention of the starving third-world farmer and community. The invention of the immediately outdated computer is the invention of toxic computer waste in garbage dumps of the Global South. The individualized automobile as the major mode of transportation is the invention of a major emitter of greenhouse gases.

Even many so-called green solutions designed to palliate the unequal global distribution of resources are governed by the unidirectional pattern of thought that characterized the development of the atomic bomb. The technological manipulation of land that defined the green revolution in the Punjab, for example, is symptomatic of an "overexposing" of land that leads to the accident of the "asphyxiation" of that land in excess salinity and water logging. Undertaken by the American biologist Norman Borlaug and funded by the Ford administration, the revolution led to the "accident" of environmental destruction and ultimately mass hunger. In other regions around the globe, similar pushes to "monocrop" by transnational corporations leads to an overexposure of the land and turns food into mainly a commodity to be sold on the international market. Monocropping asphyxiates that land and its community: it eliminates the chances of a community to fall back on locally produced food and in turn forces the community to pay for food produced elsewhere. When an "accidental" dip occurs in the market, agricultural workers' pay drops and communities already underpaid find themselves in the position of having no money to buy food. The exacerbated struggle to survive produces the "threat of encroachment" by these communities looking for new sources of food in forests and privatized farmland. This in turn requires the transformation of nature into "threatened" nature, which in turn produces legislation that represents nature as a being in need of "environmental protection," thus handing ecosystems over to the wardens of security.

As a theoretical term that troubles its traditional references, "accident" serves as a useful tool for reconsidering current political constitutions of "total war" and the way these constitutions encourage inflexible patterns of thought that position the environment as an object of investigation and use. In terms of the genealogy I've been briefly tracing here, accidents appear as events that can no longer be considered isolated or unconnected occurrences. Moreover, in a historical moment when the environment is subjected to transformations brought about by overexposure, what also undergoes transformation is the very event of transcendence. In a "totally discovered and overexposed habitat"—a strangulating version of "the end of history"—transcendence from the disciplinary confines of context becomes increasingly unavailable. As a viable activity, the idea of passing beyond the limits of a system falls out of consciousness and is replaced by the different idea of a "breakage of/by the system." Transcendence, in other words, is replaced by *Accidence*—forms of momentary breakdown that only occur *within and because of the system*, not breakdowns that radically open a free space for the potential to restructure the system. In the ontology of the Accident, productive and successful human subjects are constituted in terms of their ability, and consent, to "protect" the environment (in other words, securing the highest yield of either destructive or life-giving energy). For unproductive subjects and the poorest of humans, especially those working in the worst possible conditions in agricultural communities in the third world—the shadow humans that make the North and the West possible—they are increasingly seen as a threat to the environment and to this newly enclosed environmentality. The Accidental thus names a world in which constituents and their environments can only appear in the form of accidents: the environmentally dispossessed by capital-technological development projects, the community overexposed to toxic fallout from the "discoveries" activated by the military industrial complex, and the climate change refuge.

We can even pinpoint the specific date of the trumping of transcendence on a planetary scale by the Accidental.[9] On November 1, 1952, the H-bomb "Ivy Mike" was dropped on the island of Eugelab in the Marshall Islands of the Pacific. An operation conducted by the U.S. Atomic Energy Commission and the Department of Defense, "Operation Ivy" was the world's first successful test of a nuclear weapon. According to Elizabeth DeLoughrey, at ten megatons—"seven hundred times the explosive force of the bomb dropped on Hiroshima"—Mike "blew the island of Eugelab out of existence."[10] Its radioactive fallout "was measured in rain over Japan, in Indian aircraft, and in the atmosphere over US and Europe." Two years later the United States dropped "Castle Bravo," which increased this yield to fifteen megatons. A scientist working for the Atomic Energy Commission declared in 1954 that after just two years of testing, all humans on the planet now had "hot" strontium in their bones and teeth and "hot" iodine in their thyroid glands: "Nobody believed you could contaminate the world from one spot. It was like Columbus when no one believed the world was round."[11] As DeLoughrey reminds us, it was because of these post–World War II Cold War nuclear experiments conducted by the U.S. military industrial complex that ecology became a field of study in the university system. The Atomic Energy Commission created the discipline of "radiation ecology" to study the effects of this militarized radioactive contamination on the environment. The establishment of this field began with the hundreds of nuclear tests conducted in what was known as the "Pacific Proving Grounds." By 1958, the United States, the United Kingdom, and the USSR had exploded "nearly one hundred nuclear weapons, leading to record levels in strontium-90 in American soil, wheat, and milk."[12] Similar tests were conducted by the United Kingdom, France, and the Soviet Union and later by the People's Republic of China, Pakistan, and India. The United States conducted more tests than any other nation (see Figure 1). This Cold War arms race was, more than anything else, a race for the total control of the planet through targeting, and this race for total control generated the accident of total exposure and the impossibility for any human to transcend this contamination.

In this erasure of the transcendental in favor of the Accidental, our access to ecology changes fundamentally. *Environmentality* is the term I have been using to mark that change, and I have defined it thus far in relation to Foucault's conception of governmentality—that is, as a political constitution and administration of planetary ecosystems on the basis of their ability to be technologically "improved" so as to produce "high yields" of both consumption (more crops to feed people or more energy for fuel and electricity) and destruction (the collateral side effect of overexposed land). To this we can now add the force of the Accidental, which chokes out the possibility of alternative ecological relations for purposes

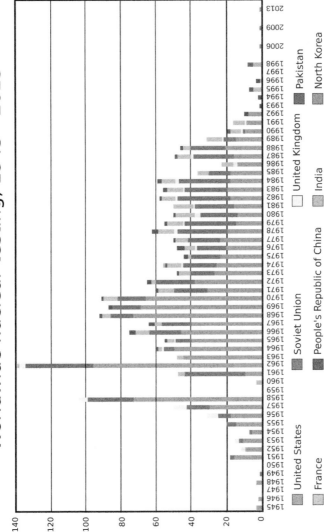

Worldwide nuclear testing, 1945 - 2013

Legend: United States, Soviet Union, United Kingdom, Pakistan, France, People's Republic of China, India, North Korea

Figure 1. This graph shows the number of atomic and nuclear bombs dropped by countries across the planet. Image courtesy of the Comprehensive Nuclear-Test-Ban Treaty Organization (CTBTO).

of total control and total security. Between 1945 and 1992, the Atomic Energy Commission and the Department of Defense conducted 1,032 nuclear tests, exploding atomic and nuclear bombs on land, in the air, underground, underwater, and in lower outer space—essentially engineering accidental radioactive fallout on a planetary scale. According to the United Nations, "Each nuclear test resulted in unrestrained release into the environment of substantial quantities of radioactive materials, which were widely dispersed in the atmosphere and deposited everywhere on the Earth's surface."[13] Nuclear testing began in the United States in July 1945. In reaction, the Soviet Union began testing in 1949, the United Kingdom in 1952, France in 1960, and China in 1964. The National Resources Defense Council estimates "a total yield of all nuclear tests conducted between 1945 and 1980 at 510 megatons . . . equivalent to over 29,000 Hiroshima size bombs."[14] Tellingly, the demand for "high yields" in environmental overproduction bears an uncanny resemblance to the toxic fallout of radiation produced accidentally in the demand to yield high destructive power from the atom. Greenhouse gases—carbon dioxide, methane, nitrous oxide—like toxic radiation are a phenomenon that cannot easily be overcome, at least for generations to come. According to a 2009 study by Susan Solomon, Gian-Kasper Plattner, Rteo Knutti, and Pierre Friedlingstein, carbon dioxide concentration "is largely irreversible for 1,000 years after emissions stop."[15]

In this sense, the Accidental that defines the perverse logic of environmentality should be understood precisely as being *systemic* in nature, not occasional, and not the result of an individual's or group's actions. It is exactly this point, for instance, that Žižek makes in his discussion of today's capitalism and its systemic, destructive power. Though he is not fully articulating the Accidental as I have been developing that concept (nor is he, for that matter, speaking about ecological concerns), Žižek nevertheless realizes the importance of what he articulates to be the "fundamental systemic violence" operative at the heart of today's neoliberal-oriented politics: "The fundamental systemic violence of capitalism . . . is much more uncanny than direct pre-capitalist socio-ideological violence: this violence is no longer attributable to concrete individuals and their 'evil' intentions; it is purely 'objective,' systemic, and anonymous."[16]

This systemic nature of the Accidental, first manifested in the early Cold War "nuclear period" of environmentality, informs the most recent, "post-9/11 period" in which climate change is fully adopted as a central military concern. In this last stage of environmentality's development, the central security gesture of the nation-state becomes a move that paradoxically strikes out against its own environmental support system. According to *Scientific American*, military and national security experts began to press scientists to provide them with a different *kind* of information about climate change in 2010.[17] Ostensibly this demand arose from the military's desire to acquire the "facts of the situation." But the demand also came with the proviso that scientific experts needed to address a weakness in their style of reporting these facts. Science, military officials argued, too often failed to emphasize *all possible outcomes* of a situation, specifically the most calamitous. In their request for more detailed analyses, the military began to push scientists to produce worst-case scenario narratives. Security experts critiqued the tendency of scientists to rely on "averages": "Military and national security experts said climate forecasters often focus on averages, or the most likely scenario, without determining the probability of an extreme climate shift."[18] This demand for a worst-case scenario is given greater weight as science shifts from a more open, intellectual community to a more militarized community. A specific issue such as rising sea levels becomes a new target for this ecomilitary machine. For instance, the Intergovernmental Panel on Climate Change (IPCC) predicted in its 2007 report that sea levels could rise anywhere from seven to twenty-three inches by 2100 but also made it clear that this future is unpredictable and thus highly uncertain. The report did not even attempt to offer an estimate of sea levels if a more rapid melting of the massive ice sheets in Greenland and Antarctica were to occur, though it alluded to the possibility. However, Jay Gulledge, senior scientist for the Pew Center for Global Climate Change (and a fellow at the Center for a New American Security), treats this uncertain future differently: "We're at a point now where we can't say we're not going to get 2 meters of sea level rise by 2100, though it's perhaps not likely."[19] Even though the worst-case scenario is unlikely, it becomes the target—the smoking gun—that confirms the need for the predominance of military modes of thought, experimentation, and action. National security experts play on the catastrophic to criticize a more patient science for its adherence to the most "likely" scenario and for its failure to give us the truth of extreme climate changes. This teleological structuring of the environment toward the goal of the catastrophic justifies the Accidental drive for the next impending breakdown, which in turn generates a tautological rationality for suspending the law. This push for the greatest possible accident therefore should not be dismissed as an empirical desire to prepare for a potential future threat to the United States. It is a policing action that deterritorializes the complexity and diversity of the environment and reterritorializes the earth so as to enframe it as a single naturalized identity: the ultimate threat to national security.

Manipulation can thus be addressed more critically (i.e., outside dominant theories of national sovereignty) in terms of the systemic lines being drawn between corporate/state flows of power and ecosystemic developments—namely, the economic, judicial, and legislative alliances being formed around the struggle for diminishing resources and the impending shift to nonpetroleum sources for global mass transportation and global mass consumption. Though the state is part of this shift, it and the cultural socius lag behind the direct activity of what should be understood properly as the developing Accidental state—a structure composed of war-oriented environmental organizations, state and corporate officials, security institutes, centers for foreign and domestic policy, private companies, and the military. As I argue in the next section, in the "Accidental state" humans and the environment—biopower and ecopower—are situated as mutually exclusive problems to a political regime caught within the problematic of adaptation. When one problem (humans) enters the domain of another (an ecosystem), the "solution" can only be a destructive one: the "accident" of encroachment and subsequent "accident" of forced eviction, which more often than not lead to increasingly violent encounters. The logic of adaptation thus utterly depends on the belief that environmental relations, specifically in terms of the struggle for diminishing resources, will become increasingly antagonistic.

The Metaphysics of the Accidental: Transcendental Mapping and Imaging

The desire for the total destruction of the enemy's territory that characterized World War II and Cold War developments in bombing technology goes hand in hand with an earlier military interest: the need to acquire absolute knowledge of enemy territory, which evolved into the fantasy of "light everything up" symbolized in the imperial search for the "totally discovered planetary habitat." As we saw in the introduction, President Eisenhower's establishment of the International Geophysical Year (IGY) in 1957 brought within scientific reach the dream of mapping an entire planetary ecosystem. But this fantasy of total surveillance was not only a dream of the Cold War. It manifested itself much earlier in World War I experiments in aerial photography, which were in turn influenced by even earlier imperial eighteenth- and nineteenth-century practices in national and colonial mapmaking. As I will explore here and in subsequent chapters, the belief in "total mapping" as a step toward acquiring absolute knowledge is one of the elements of environmentality's intelligibility and a central demand of the security society. The obsession with absolute knowledge is made possible by a certain relation to transcendence. But to fully comprehend the nature of this obsession, one needs to understand environmentality's relationship to modernist conceptions of universality and particularity.

Adaptation, for instance, presents itself as a universal answer—one acquired after gaining total knowledge of a situation. After examining all the variables, it becomes clear that adaptation is the best solution. Everyone needs to adapt; even nature itself is defined by neo-Darwinian "security biologists" in terms of its organic processes of adaptation. The possibility that adaptation might be related to particular interests therefore never enters into the picture. But contemporary theorists such as Judith Butler, Ernesto Laclau, and Slavoj Žižek argue that the so-called universal position is in fact an empty space without any empirically verifiable content. They read the supposed universal solution as nothing more than a power takeover of this empty space by a *particular* point of view that has successfully hidden its special interests. To phrase this specific relation in Hegelian terms, the transcendence of a particular position (developing the bomb before Germany, adapting to impending climate change) to the *only viable* position (we *had to* drop the bomb, adaptation is the *only realistic solution*) universalizes the particular. But this universalizing occurs in such a way that the originary, empty space of the universal is forced into becoming a "concrete universal."[20] The making concrete of the open nature of the universal amounts to the forced minimization of other points of view into one all-encompassing order. This act of colonizing the empty space of the universal by a particular leads to the trumping of transcendence by Accidence.

We can understand this takeover of the universal and the desire for absolute knowledge in specific ecological terms by considering how the universalization of a single methodology can adversely impact the specifically particular and heterogeneous nature of ecological relations. For instance, the exalted idea of technological progress, as many environmentalists and ecocritics (Commoner, Rodman, Capra, Merchant) have pointed out, has from its beginnings in modernity been distinguished by a mode of disinterested inquiry indifferent to the relational nature of ecosystems. However, not widely acknowledged or examined is the inherent connection of this disinterested inquiry to the security society's search for a presumably organic, inexorable movement of contemplation. In the twentieth century, the idea of technological progress would be marked by the incorporation of a certain unidirectional scientific approach into the military industrial complex. Disciplines—namely, certain formations of physics and biology and later information technology, computer science, and game theory—came to grow in line with a pattern of experimentation that was analytical, linear, and strategic in nature (as opposed to the systemic and holistic approach favored by environmentalists).

As I mentioned briefly in my introduction, in biology (a discipline central to both ecology and the military), the ontological adherence to a discrete, linear analysis generated the field of *molecular* biology, which began during the Cold War to supersede *classical* biology. Classical biology emphasizes the interrelational "complexity of the whole cell." It depends on the intellectual capacity to consider the ecosystemic effects of biological chemistry.[21] It focalizes contextual—that is, *particular*—relations and never assumes that a single element within a whole can be known absolutely or that it ever operates universally in determining the function of a whole cell. Molecular biology, on the other hand, treats chemical constituents as separate from the cell; it is what we might call "target oriented" in that it specifically targets certain constituents as more important in order to discover inherent and usable properties. The environmentalist and biologist Barry Commoner criticized this split when it began to achieve prominence at the height of the Cold War. He tacitly associated it with Cold War politics but did not comprehensively theorize the connection between the two. Instead, he emphasized the failure of molecular biology (which was favored by the military industrial complex) to consider its relation to varying contexts. The development of this split began to erase concerns about environmental interactions. New chemicals produced by the military and subsequently introduced into detergents in the early 1960s, for instance, resulted in heavy pollution of rivers with detergents; the chemical compound failed to break down in the environment, and the widespread problem demanded national legislation. Similarly, the introduction of a compound of DDT in a Bolivian town meant to destroy malarial mosquitos also killed the local cat population: "With the cats gone, the village was invaded by a wild, mouselike animal that carried black typhus. Before new cats were brought to restore the balance, several hundred villagers were killed by the disease."[22]

To return to the context of the race to develop the bomb and its relation to absolute knowledge, as Paul Virilio shows in an earlier work on military history, *War and Cinema: The Logistics of Perception*, transcendent techniques of military activity such as aerial bombing grew as a secondary development in reaction to the primary demand for establishing a system of total, universal surveillance during World War I. The "supply of images" that was generated in the process of gaining information about the enemy's geography served as the first "ammunition supply." The idea that planes could also drop bombs arose only as an afterthought of the primary desire to discover a systematic machinery of intelligence gathering that would enable the military to gain absolute security of enemy territories.[23] In other words, the explosion of the bomb/insecurity does not happen first; it does not preexist the securitizing surveillance system. That system, that pattern of thought, produces the guarantee of destruction.

These technological developments, generated and overseen by the war machine, would simultaneously embed into the activity of representation the fantasy of absolute, conclusive knowledge and the desire for decisive destruction. Such efforts inform the unidirectional and universalizing pattern of thought operative in the discourse of environmentality. According to both Virilio- and Deleuzian-influenced historical theorist Manuel De Landa, from World War I on, representation would be coconstituted alongside the need for intelligence gathering (image capturing) and targeting. (As I will argue in chapter 2, the conjoining of these two within the representative act exists in a latent form in the nationwide surveillance and documentation apparatus of the enclosure act.) The increased demand in the precise knowledge of the enemy's geography, the need to "light everything up"—which generated a generalized "total comprehension" of "intelligence" that supported a military point of view—was made possible in technology by the integration of the combat vehicle and the camera into a new form of optical "weapons system."[24] The development of "sight machines" loaded onto planes for precision aerial photography began to spawn a naturalized form of "automated perception." Automated electronic search systems began to replace human eyesight, which was now presumed to be limited and incapable of handling the war machine's desire to control vast environments. This automatic perception in turn would also begin to generate an "automated interpretation of reality," for the information accumulated would serve as the database for the command center and its qualified personnel. As a military "representation machine," the camera would develop into a major vehicle for data accumulation in the years leading up to and following World War I, erasing multiple and contextual perspectives with the concrete universal of a "higher," totalizing point of view. As De Landa has argued, cinematography and the cinema system would unfold out of these same developments in manufacturing "flying observation posts."[25] The "magic" of early movie making would be tied to military attempts to provide a "complete and total picture" of an enemy's territory. The cancellation of space and time in the totalizing "view from above" would match the "spellbinding" nature of being in the movie theater, which came to be such a popular entertainment phenomenon because it was "capable of giving the audience, in every fraction of a second, that strange sensation of four-dimensional omnipresence cancelling time and space."[26] Judith Butler, in her discussion of the first Gulf War in Iraq, reveals how this metaphysics of divine positionality results in the "conflation of the television screen" with "the lens of the bomber pilot."[27] The televised representation of the war, viewed by Americans thousands of miles away from its immense destructive violence, becomes an extension of the targeting apparatus of the war machine. Its depictions, taken as disinterested visions of the media rising above partisan interests, are not an unbiased "*reflection* on the war," Butler argues, "but the enactment of its phantasmatic structure."

This captivation with a transcendent view from above as the perspectival quintessence of representation that domesticates the elusive movements and geographical partialities of time and space has, of course, a long Western and imperial history. The transcendental viewpoint was naturalized repeatedly, for instance, in and through the influential cultural site of literary production. The "prospect view" poets of the seventeenth and eighteenth centuries, for example, along with novelists such as Daniel Defoe, Henry Fielding, Tobias Smollett, and others (and later James Fenimore Cooper and painters of the Hudson River School in America), popularized the idea of a totalizing view from above that transcended the limitations of space and time. The ideal of the prospect view achieved its highest formation in the nature poetry of Wordsworth, who "sought to create a map of the nation that transcended local differences by subjecting the landscape to the 'prospect view,' the abstract imperial gaze available from the mountain summit."[28]

Wordsworth's transcendental literary mappings paralleled the contemporary scientific, phantasmatic "national mapping" project in England conducted by the military during the Napoleonic Wars. In preparation for a potential French invasion, the office of Ordnance Survey (OS)—which had been formed at the request of King George II to survey the Scottish highlands after the Jacobite revolt of 1745 and would later become part of the Ministry of Defense—began an extensive, sixty-year survey of England, Wales, and Ireland. It may not have had the precision of satellite image making, but it was the eighteenth century's version of Eisenhower's IGY. The project was known as the "Principle Triangulation of Great Britain" and would become the basis of a new ground of transcendental accuracy—a politico-scientific demand inaugurated by the extensive enclosure movement surveys, which would later be exhaustively strung out in the twentieth century with satellite technology. With developments in printing technology and telecommunications in the nineteenth and early twentieth centuries, maps would become a standard and influential feature of newspapers such as the *Times* (of London), the *New York Times* (which began to regularly print maps during the Civil War), the *Wall Street Journal*, and the *Christian Science Monitor* (the *Journal* and the *Monitor* both have a history of devoting more than half their maps to military conflicts).[29] In the twentieth century, maps began to appear in newspapers daily, and the map as a scientific-phantasmatic instrument of military data accumulation and analysis would expand to become a recognized method for diagramming economic resources and developments, scientific explorations, natural disasters, transportation innovation, tourism, and a variety of geopolitical matters of security. These and other publications normalized and expanded the logical economy of the map, which functioned as a significant instrument for growing both a national and a metropolitan consciousness.

Similar mappings of the colonies began at roughly the same time as this mapping of the empire's domestic landscape. The national ideological investment in mapping as a military will to mastery over domestic and foreign landscapes, buttressed by the scientific demand for rigorous data accumulation, would become a key component of colonial expansion. The "Great Trigonometrical Survey of India," which began in 1802, sent surveyors scrambling to find "high vantage points" that coupled the transcendence of sight (as the ideal optical point of scientific accuracy) with the military "high ground" (the ideal position of defense). Cartographic control served as the basis for the "political control of all regions."[30] Surveyors were "active agents of imperialism rather than passive data collectors."[31] The survey deployed the self-validating logic of scientific accuracy to generate the fantasy of intellectual superiority, in turn based on the fantasy of establishing an infallible method of apprehending the world.[32]

In cultural terms, this representational method is a key component of Western, and later non-Western, literature. Narratives of the transcendental containment of ecologically and linguistically open or "common" spaces, for instance, are one of the novel's most predominant themes: from *Robinson Crusoe* to *Heart of Darkness* to *Midnight's Children*. The discourse of Western, instrumental metaphysics is the primary ontology informing this representational (linguistic) and spatial (geopolitical) war against the commons.[33] Here it is important to recall the twentieth-century philosophical critique of Western metaphysics, begun with Nietzsche and Heidegger and so central to the later poststructuralist work of theorists such as Derrida, Foucault, Kristeva, Spivak, Virilio, De Landa, and others.[34] The metaphysical (*meta ta physika*: thinking from a transcendent position "over"—*meta*—temporal being as such) idea of gaining mastery by rising "above," "beyond," and "outside" the heterogeneity and flux of changing historical and unenclosed ecological contexts constitutes one of the essential exercises of a colonial order.[35] The seemingly innocuous "view from above" that successfully captures and oversees immense geographical areas has a powerful effect in regulating the capacity of human consciousness to conceptualize large areas of our planet's environments. Robinson Crusoe, for instance, does not successfully colonize his island until he obtains a privileged vantage point above its "wildness" by physically and ideologically transforming the open space of the land into an "improved" and "cultivated" enclosed fiefdom.

Rudyard Kipling's popular novel *Kim* is perhaps the best example of the colonial naturalization of metaphysical observation and its coupling to the discourse of enclosure. The empire's conversion of Kim into a soldier capable of using his vision to map, target, and classify the "great whirl" of India from a superior height above (he becomes a mapmaker for the British Imperial Indian Survey) represents one of the key twentieth-century goals of the Western military machine: the successful development of what modern-day military authorities would begin to refer to, in relation to the science of governing soldiers, as a "higher-order motor." (As we will see in subsequent chapters, the higher-order motor is the name for a formal mechanism that attempts to seamlessly combine the elements of command, control, communication, and information—known in scientific-military discourse as C3i, or C4i when "computers" are added to the equation.)[36] Kim is trained by military authorities to "map" India with his eyes: "Thou must learn how to make pictures of roads and mountains and rivers—to carry these pictures in thine eye till a suitable time comes to set them upon paper."[37] Kipling resolves the tensions surrounding Kim's struggle to find both a spiritual meaning in life and an identity proper to the "great game" of empire by having Kim's teacher—the lama—offer up a particular form of visual transcendence as the ultimate solution to the question of existence. At the conclusion of the novel, the lama invokes the phantasm of Matthew Arnold's imperial edict to see all life "steadily and whole," thus rising above the commons and its constant differential and profane transformations: "My soul grew near to the Great Soul which is beyond all things. At that point, exalted in contemplation, I saw all Hind, from Ceylon in the sea to the hills . . . I saw every camp and village, to the least. . . . I saw them at one time and in one place, for they are within the Soul. By this I knew the Soul has passed beyond the illusion of Time and Space and of Things. By this I knew I was free. . . . So thus the Search is ended."[38] Kim only achieves his full development and "true identity" when he learns to rise above the chaotic and undisciplined commons of India and when he discovers how to think "rationally" by rising above the linguistic commons to use the imperial language of proper English to compartmentalize the ecosystem below into classifications useful to the expansion of the colonial empire. These maneuvers rationalize a certain apprehending of the world, make possible the colonization of immense spaces, and authorize a presumably superior basis for thought itself.

The System of the Sages

The view from above is thus ontologically laced with superiority, capturing, control, manipulation, targeting, predation, the erasure of distance and difference, the spatialization of time, historical collapse, empirical knowing, and a host of other metaphysical concepts and strategies. The ecological and linguistic status of the reterritorialization of open fields by the enclosure movement (which continues to this day in the Global South) hinges on the establishment of a representational apparatus that serves as the unquestioned metapolitical and cosmopolitan norm. In the post–World War II era, the search for a metapolitical "higher-order" apparatus expanded rapidly across scientific and military registers of knowledge production. This search for a higher-order mechanism eventually led to the constitution of the twenty-first-century form of environmentality that seeks to manage climate change and can be found operating in security biology's search for an organic architecture of "natural security."

The essential metaphysical relationship between the presumably impartial, scientific "gathering of information" and the militarized view from above is perhaps best visualized in the Air Force massive surveillance program known as SAGE (Semi-Automatic Ground Environment)—the first and most ambitious of the early–Cold War security society innovations. The SAGE system grew out of the initial "revolution in military affairs," which was renewed by CIA Director R. James Woolsey after the collapse of the Soviet Union. American computer and systems dynamics engineer Jay Forrester was one of the men at the forefront of the SAGE project, emphasizing the "need to oversee and operate large complex organizations."[39] According to historian Paul N. Edwards, SAGE (and its predecessor, MIT's Whirlwind computer) was "by almost any measure—scale, expense, technical complexity, or influence on future developments—the single most important computer project of the postwar decade."[40] The program's goal was to establish a "continental fortress" above and across the entire geography of the continental United States (see Figure 2). Essentially a massive computer designed for purposes of panoptic surveillance, the SAGE project was overseen by the Air Force and developed at MIT, where "the general of physics"[41] Vannevar Bush (engineer and former chair of the National Defense Research Committee during World War II) had established the Office of Scientific Research and Development (OSRD) for President Roosevelt. Bush almost single-handedly orchestrated the embedding of the security society at the academic level. As director of the OSRD, he oversaw the efforts of six thousand American scientists engaged in the goal of applying science, technological development, information gathering, and image analysis to warfare.

SAGE was housed in MIT's Lincoln Labs (the corporate-university workshop of MIT, IBM, and Bell Labs) and exceeded the Manhattan Project in funding and scale. The project came into existence through the

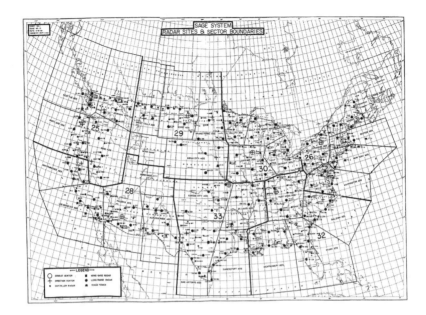

Figure 2. A map of the SAGE "continental fortress," showing the range of U.S. continental security surveillance.

establishing of a standard commerce between four corporate-university contractors: IBM invented the computer hardware, the Burroughs Corporation generated a communications array, Western Electric designed and constructed SAGE's twenty-three "Direction Center" buildings, and Lincoln Labs provided the necessary system integration. Twenty-two interconnected "SAGE Centers" were constructed across the United States and one in Canada. SAGE was the largest and, in its day, the most advanced computer system in existence.[42] Tellingly, as Edwards reveals, the SAGE system actually had little defensive potential: its directional centers were poorly placed; its tests under combat conditions revealed many flaws; and by the time it was fully operational, it would have been easily targeted and destroyed by Russian intercontinental ballistic missiles. However, the research that went into its development was instrumental in generating the new industry of militarized and computerized security. And more important, its impact as an idea helped spread the fantasy of environmentality—the imagining of the planet as a securely enclosed and militarized environment—to all levels of the newly forming national security state:

SAGE was the first large-scale, computerized command, control, and communications system. . . . [I]t unleashed a cascading wave of command-control projects from the late 1950s onwards, tied largely to nuclear early warning systems. These systems eventually formed the core of a worldwide satellite, sensor, and communications web that would allow global oversight and instantaneous military response. Enframing the globe, this web formed the technological infrastructure of closed-world politics. . . . Beginning with SAGE, the hope of enclosing the awesome chaos of modern warfare (not only nuclear but "conventional") within the bubble worlds of automatic, rationalized systems spread rapidly throughout the military, as the shift to high-technology armed forces took hold in earnest.[43]

SAGE led to developments such as online systems, personal computing, modems, interactive graphic displays, and the computer mouse. The program spawned the professions of software development engineering and programming and institutionalized the field of data analysis. IBM, at the time a little-known company, came to dominate the computer industry because of its work on SAGE. The SAGE program and multisite system functioned for more than twenty-five years until 1983. According to James Ray, second lieutenant and SAGE programmer at the Great Falls site from 1966 to 1969, "SAGE was one of the engineering marvels of the century. In fact if it looked better [than its unadorned cement buildings], it might have been ranked with the Eiffel Tower, the Saturn Rocket, or the Golden Gate Bridge."[44]

SAGE was designed to solve the threat of enemy bomber aircraft entering American airspace and dropping their "payload" before American planes could intercept. Because of their ability to easily penetrate homeland defenses, bomber aircraft were considered to be absolutely immune to any defensive strategy. A successful defense against their approach would require thousands of interceptor planes to be in the air around the clock—a technological and economic impossibility. In response to this Cold War tale of full enemy attack, the military installed devices that could link together radar installations around the United States, thereby generating a technological apparatus that could analyze signals twenty-four hours a day. In turn these signals would form an early warning system, sent directly to pilots constantly on call to fly interceptor jets. SAGE was

an idea forged out of the demand for a total national defensive carapace—
the dream of a miraculous device that would monitor the entire domain
of an ecosystem. It functioned by fully interlocking the disciplined soldier with technology so as to create a physical apparatus of absolute speed:
"Both the engineering and the politics of closed-world discourse centered around problems of *human-machine integration*: building weapons,
systems, and strategies whose human and machine components could
function as a seamless web, even on the global scales and in the vastly
compressed time frames of superpower nuclear war."[45] SAGE was designed
to be "a total system, one whose human components were fully integrated
into the system."[46] The result was a generalized technohuman consciousness that would reach from the various command centers distributed
throughout the United States to the entire line of the nation's geographical
borders—a technologically verifiable version of Benedict Anderson's linguistically imagined community developed by the physics and engineering
industries and run by the war department. (This same strategic mandate
to "fully integrate" was repeated in the Center for Naval Analysis's *National
Security and the Threat of Climate Change* report, which ended with a series
of recommendations calling for the phenomenon of climate change to be
"fully integrated into national security and national defense strategies."[47])

More than the creation of an advanced system of total surveillance, the
SAGE system made manifest (and cemented in technological materiality)
a certain dream of combining and managing the scientific administration of enclosing, mapping, image making, and observing—fueled by the
tactical demand for universalized forms of speed and power: "Cold
War military forces took on the character of systems, increasingly integrated through centralized control as the speed and scale of nuclear war
experiences, linked the globalist, hegemonic aims of post-World War II
American foreign policy with a high-technology military strategy, an
ideology of apocalyptic struggle, and a language of integrated systems."[48]
The success of standardized and advanced systems of targeting, coupled
with the destructive firepower of atomic and nuclear weapons meant that
SAGE could saturate the unidirectional force of the Manhattan Project at
an unprecedented level across the nation-state: "national security" had
entered its adulthood.

The SAGE system hardwired the idea and ideal of security into the performative identity of the neomilitary United States. The new "narration of
the nation" no longer had a past, present, or future; its temporal character

became the intensified and asphyxiating totalized present, a present forced to be in a suspended state of alert for the next spectacular threat. It was a mode of production designed to impose tighter controls on existence, transform images into powerful weapons of a security-state arsenal, and, in the interpellation of public consciousness to focus on such sensation-driven images, productively foreclose alternatives. As an immense amount of recent critical American scholarship has argued, this governing concern for security would eventually expand from the military to the civilian sector, reaching an apex of heightened spectacle-oriented defensiveness in the post-9/11 George W. Bush era of what Giorgio Agamben has identified as the "exceptional" state of existence—a social order in which the law and civil existence are "necessarily" suspended.

Spotting and targeting the air commons above was also understood to be a form of saving the land below, which took on the new meaning as an always-threatened ecosystem (see Figure 3). Later developments in satellite technology would invert this view of the skies to the prospect-view performance of spotting and targeting the enemy on the ground from above (first embodied in the IGY and its inauguration of satellite technology). The manifestation of the military provenance of the view from above in Kipling's novel thus expands into more technological modes of production with the developments in information and data gathering in the second half of the twentieth century—methods that privilege and combine scientific accuracy and security with optic transcendence. In the development of the SAGE optics system, the "Charactron" was one of the first of a new breed of security technologies that brought the centuries-old dream of a totalized metaphysical system of control to scientific completion.

The Charactron acquired its name through the action of pinpointing certain targets moving in the air and on land and recoding these targets into letter characters. This enables the targets to be transformed into electronic words and phrases that can be decoded by human observers—a methodology that matches the discrete targeting of molecular biology. Figure 3 shows the capturing of the earth's ecosystem into this mechanical, calculative language of military security. The figuration of the "electron optics" advertisement from 1962 celebrates the achievement of a secure, transcendent perspective in its aesthetic representation of the actual technology. In the fictionalized scenario of the advertisement, the screen of the control panel is transported to a divine height above the surface of the earth, indicating its ability to surmount geographical

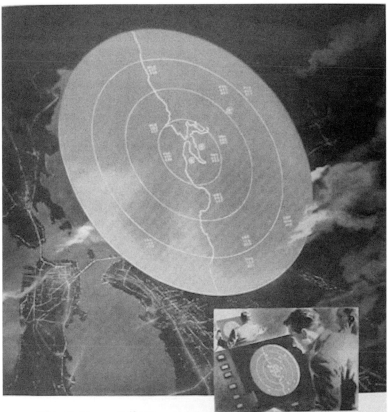

Electron optics

In *Charactron®*, as shown above, the principles of a cathode ray tube have been refined with pinpoint accuracy and applied to the presentation of aircraft surveillance data in "SAGE" . . . the Continental Defense System.

The precision required in manufacture is of the highest order.

Heart of the tube is a circular matrix on the face of which, in a ¼" square, 64 minute code characters have been punched. The electronic beam, res-

ponding only to processed radar data, selects these characters individually, at speeds up to 10,000 a second—and displays them in groups, identifying targets by type, speed, altitude and track number.

This new tool, the *Charactron®* electron optic tube, has many applications besides that of air surveillance. We'd welcome the opportunity to consider any problem of yours which this latest development of precision work in electron optics might well help to solve.

STROMBERG-CARLSON COMPANY S-C

A DIVISION OF GENERAL DYNAMICS CORPORATION

General Offices at ROCHESTER 3, N. Y.

Figure 3. A 1962 advertisement for the SAGE "Charactron" tracking device.

distance and overcome the "drawbacks" of temporal displacement, geographical distances, and the general unsecured chaos of the earth's unruly
ecosystems. The narrative below the artist's rendering explains that this
floating electronic battleship of the SAGE Continental Defense System
functions optimally—with "pinpoint accuracy." The optic mapping armature has the power to collect data by (re)visualizing any environmental
context—piercing through the darkness of night and the cover of shifting, cloud-pervaded patterns of weather. In the smaller image, the trained
air force officer leans over a representation of the actual Charactron surveillance panel, reading and interpreting the complex and heterogeneous
ecology of the planet through the scrutinizing process of what we might
call "security reduction": the collapse of the diversity and flux of ecological life on the ground into immobilized and essentialized identities
easily categorized and targeted. The language of the Charactron narrative
confirms the metaphysics of reducing all life to identifiable, self-present
identities: "The electronic beam, responding only to processed radar data,
selects these characters individually, at speeds up to 10,000 a second—and
displays them in groups, identifying targets by type, speed, altitude and
track number." The entire process of control is clearly gender specific: the
calculating male soldier overseeing this operation, the image indicates, is
part of a network that extends to other male soldier data opticians and
male soldiers relaying, on the phone, targeting information to the central
controlling site of the "command center."[49] The military-technological and
political dream of total metaphysical control comes at the cost of reducing the ecological differential interaction of habitations to identifiable and
assailable targets.

Al Gore's success at revising the MEDEA program in 2010 suggests that
such methods of ecological representation may continue. Figure 4 shows a
formerly classified CIA satellite image released to MEDEA scientists in 2010.
This prospect view successfully indicates ecological changes in the status of
ice at the North Pole—information certainly crucial for understanding our
environmental crisis. The use of this information is another question entirely.
Even in the apparent declassification of this metaphysical data, scientists
must deal with images purposefully "degraded" by the CIA so that the security institution can "hide [its] satellite's true capabilities."[50] If we are supposed
to see a connection to Euripides's play *Medea* in this astonishing acronym,
are we meant as well to see the face of an untrustworthy mother in the
melting ice? Conversely, in the insert—another transcendental-Accidental

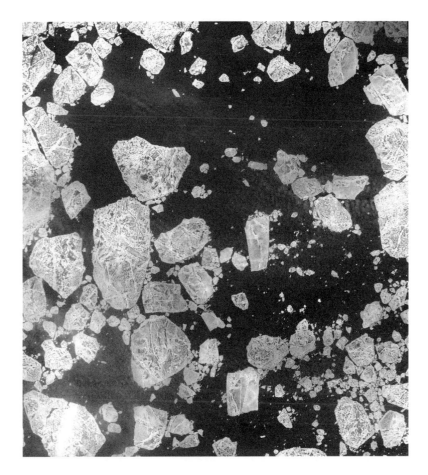

Figure 4. A representation of a CIA satellite image released to scientists working in the MEDEA group. Image courtesy of the U.S. Geological Survey Global Fiducials Program.

mapping of the Pole (I refer here to the first public appearance of this image in the New York Times, which included an insert map of the North Pole, showing the loss of total ice sheets)—are we meant to see the wisdom of the SAGEs, giving us the fantasy of an unbroken icecap? If the one is security and the other insecurity, then where lies our access to alternative ecologies?

In Euripides's *play*, Medea is abandoned by her husband Jason. He leaves her after ten years of a happy marriage when King Creon offers him his daughter Glauce. Jason had been a well-known and loved hero for years, successfully performing a series of grueling tasks—all with the significant help of Medea. In the play Medea takes revenge on Jason by killing her children. He sends Glauce a poisonous dress and coronet, resulting in the deaths of both Glauce and her father. But the story of filicide is Euripides's peculiar invention. Earlier stories of the Jason and Medea myth speak of the children's death either as an accident or as the result of an organized attack by the citizens of Corinth. Nonetheless, Medea is handed down by history as little more than an untrustworthy enchantress and witch. In the early years of the Cold War, a group of scientists with expertise in nuclear technology began to offer their recommendations to President Eisenhower's military leaders. They began to refer to themselves as the Jasons, and though they were civilians, they worked directly with the Department of Defense.[51] In this revival of the ancient granddaughter of the sun god Helios, are we to understand this new version of MEDEA to be an invaluable companion or an untrustworthy enemy? Perhaps we will be made to think that the earth itself is Medea, a force in the midst of killing its own children. In this sense, we would then have to see the new, post-9/11 military advance against global warming as a revival of the Jasons. As the security society expands its deployment of insecurity to the event of climate change, it becomes increasingly difficult to tell precisely who is unfurling the toxic coronet and dress and who is being targeted with the blame.

The purity of the cold, calculating character of electron optics; the self-evident solution of the atomic bomb; the practical necessity of adaptation; the Cold War interest in manipulating the environment on a planetary scale; and the mirage of absolute, accurate knowledge obtained through metaphysical surveillance—all of these, as I've been arguing, are essentially related. And all are increasingly centered and orchestrated around the global accident, the ground zero of environmental decline. If the projections of the IPCC are correct and successfully hailed within the fantasy of the security society, the mandate to guard against ecological alternatives from "crowding out full consideration of adaptation challenges" will more than likely evolve into a generalized fever—Jason and the Sages of Corinth against Medea.

Whatever the possible outcome, the war games of this future will be ecological war games—maneuvers zeroing in on shifting climatic zones, shrunken and desiccated ecosystems, lost land, unstable borders—with "ecopower" lancing biopower and environmental threats exceeding the former threats of communism, rogue states, and terrorism. This annexation of ecological futures to the logic of security begs the question, what might alternatives to these militarized maneuvers look like? What might a future other than an adaptation to the technological, enclosing legacy of climate change be? I move toward a tentative conclusion in this chapter by considering these questions alongside two recent theoretical engagements. I then situate those engagements in relation to a specific political struggle over an environment threatened by the metaphysical policies of neoliberal development.

Spectacularizing Accidents: Visualizing Environmentality

How can thought collect Debord's inheritance today, in the age of the complete triumph of the spectacle? It is evident, after all, that the spectacle is language, the very communicative and linguistic being of humans. . . . [C]apitalism . . . not only aimed at the expropriation of productive activity, but also, and above all, at the alienation of language itself, of the linguistic and communicative nature of human beings, of the logos in which Heraclitus identifies the Common. The extreme form of the expropriation of the Common is the spectacle, in other words, the politics in which we live. But this also means that what we encounter in the spectacle is our very linguistic nature inverted. For this reason (precisely because what is being expropriated is the possibility itself of a common good), the spectacle's violence is so destructive; but for the same reason, the spectacle still contains something like a positive possibility—and it is our task to use this possibility against it.[52]

Agamben's return to the "positive possibilities" of Guy Debord's theorizations of the "society of the spectacle" touch on one of the most important concerns in the struggle for the future safety of human and nonhuman inhabitants of ecosystems around the planet: the relationship between a successful political struggle that gives agency to inhabitants and the need to break through the violence of the metaphysical spectacle that, in its essential nature as a targeting device dispensing what Virilio has aptly termed "information bombs," erases the complex politics of ecosystem commons.[53] As Rob Nixon has argued, the silencing of nonspectacularized, site-specific struggles constitutes one of the key representational and strategic challenges that ecocritics and environmentalists face in today's sensation- and military-driven world.[54] Environmental struggles all too often occur off the spectacularized-accidentalized representational radar of public consciousness. When they do enter this representational order, they tend to visually represent troubled environments through the prob- lematic (in Althusser's definition of the word)[55] of a reverse racism that characterizes the struggles of indigenous populations and their habita- tions as agentless victims destined to succumb to the rapaciousness of energy-hungry corporate developers out to replace the purity of natural environments with the technological marvels of economic expansion.

One such recent example of the problems of ecological image mak- ing has arisen from the controversy surrounding the "Belo Monte dam" project in Brazil, which promises to destroy indigenous lands and inflict untold damage on the Amazon ecosystem. The Belo Monte dam is the embodiment of an asphyxiating technological cure-all generating the accident of ecological destruction. If constructed, the Belo Monte dam will be the third largest in the world. The Brazilian government and Norte Energia, the consortium overseeing the project, characterize the dam as a "clean energy" solution that will provide electricity for twenty-seven mil- lion homes and jobs to thousands of local citizens.[56] However, the project will flood approximately 250 square miles of the Amazonian rainforest— erasing the village of Santo Antonio in addition to a quarter of the city of Altamira—and almost entirely cut off the flow of a 150-mile stretch of the Xingu River. Part of an immense developmental proposal to build some sixty dams in the Amazon basin, the dam will dramatically change the Xingu River, altering and in some areas destroying the com- plex environment and biodiversity of the Xingu basin, a region that was recognized by the Brazilian government in the 1950s as the nation's first indigenous territory. The basin is home to some twenty-five thousand indigenous peoples from eighteen ethnic groups. The dam's rationale thus hinges on the absolute transformation of the incalculable heterogeneity of an eco*system*—full of highly complex particularities—into the concrete universal of an electricity factory designed to ensure energy security not for indigenous locals but for the compounds of the military, the state, and the corporate and private elite living in cities elsewhere.

In June 2012, eighteen indigenous groups organized "Xingu+23," a pro- test event that paralleled the Rio+20 Earth Summit. The "+23" refers to the first meeting of indigenous peoples twenty-three years ago, an organizational event that successfully opposed the construction of the dam back in 1989. After the protest of 1989, the dam was put on hold. But the Lula government began to pave the way for the revival of the project in 2002. In 2010, new plans started to go through the finalization process in government committees and the courts, and in 2012, under the Rousseff administration, construction of the dam began again.

For the past two years, indigenous peoples of the Xingu River basin have been organizing to protest this "clean energy" project without much interest or coverage by the global media of the North. But Earth Summit II, which took place in Rio in the summer of 2012, and the series of protests organized around the summit gave the media the global-level market capital they needed to send reporters and photographers to subsume Rio+23 within their neoliberal interpretive apparatus. The event was thus repositioned on the larger stage of planetary politics. The media quickly produced a series of images purporting to present the truth of the ecological commons of the Amazon and its indigenous peoples to audiences in the Global North and West, arresting the public's attention against the background of other spectacularized accidents of global violence.

Covering the event required the media to produce images that would capture the public's attention through the performative force of the (anticommons) spectacle. These images were loaded with the specific regulatory figurations through which the neocolonial state governs its citizens' understanding of political events unfolding in the Global South. At the same time, the actions of the indigenous peoples involved in the political struggle to maintain their centuries-long status as inhabitants of an ecological habitation introduced a different power that the state paradoxically saw but did not recognize—a power that disrupted the traditional image-making apparatus through which the media produce understandable and normative representations for the world market. So far in this chapter, I have been speaking about the rise of a particular ecological understanding brought about by military and scientific developments generated by the imperatives of the security society. As I have attempted to show, these developments are also part and parcel of a general representative apparatus, one that attempts to produce disinterested, scientific environmental knowledge by targeting landscapes from a metaphysical position of purported necessity that rises above partisan interests and alternatives. This image-making apparatus is now standardized and powerfully defines the figuration of images for the ecological commons and the global market in the twentieth and twenty-first centuries.[57]

The Amazonian rainforest is, for instance, one of the most represented and talked-about environments on the planet. It is also one of the last remaining ecological commons, increasingly under the threats of neoliberal development and forms of human encroachment brought about by various neocolonial forms of displacement. The "commons" is of course a term used widely in environmental discourse. It is generally understood as a spatial concept denoting an ecological interaction that opposes today's enclosures and privatizations of geographical locations. But as Agamben reminds us, the term also holds representational significance: it refers specifically to the ethical concept of a common good fought for on a linguistic-representational register in addition to the spatial dimension of struggles over unenclosed territories and environments. In this sense, we might define colonial orders of domination in terms of how they seek to supplant open spaces and how they seek to suppress an ethics of an open-ended representational struggle (grounded in the free space of the linguistic commons) by imposing an order that subordinates heterogeneity and the complex diversity of ecosystems and their inhabitants to a predetermined end.

The primary images produced in the wake of the Center for a New American Security's Climate Change War Game, for instance, were generated as an attempt to interpellate the multiple and contestatory U.S. political and civil communities within the militarized framework of adaptation (see Figures 5 and 6). Produced by the U.S. National Center for Atmospheric Research and dubbed the "Angry Red Future" slides, the images function as a powerful mobilizer for a generation made insecure by global warming anxiety. Again, in terms of breaking through the political impasse of the crippling debates about the "truth" of climate change, such images carry great potential. However, in the context of the genealogy I've been laying out here and in the context of the war machine, their value lies elsewhere: they are deployed to further solidify the needs of the security society, not, for instance, to challenge the developmental processes of neoliberalism that support the general apparatus of security, and not, to name another instance, to encourage humans to rethink their atomized, enclosing relations to the planet's overexposed environment. Their complicity with the fantasy of absolute security—in and through their representation

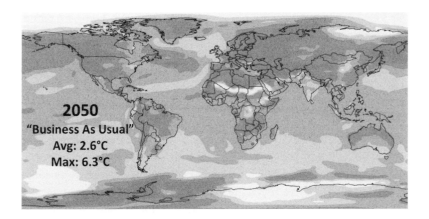

Figure 5. The Center for a New American Security's visualization of global warming, appended by the heading "2050 'Business as Usual' Avg: 2.6 degree C Max: 6.3 degree C." Image courtesy of Oak Ridge National Laboratory, U.S. Department of Energy.

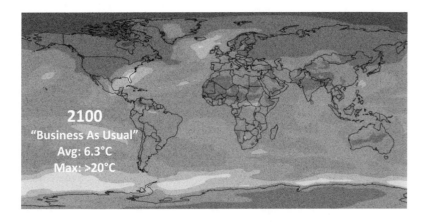

Figure 6. The Center for a New American Security's visualization of global warming, appended by the heading "2100 'Business as Usual' Avg: 6.3 degree C Max: >20 degree C."[56] Image courtesy of Oak Ridge National Laboratory, U.S. Department of Energy.

of the earth within the fantasy of absolute insecurity—makes them the consignment of the Accidental state. They are meant to function as striking visual performances of the terrifying *accident* of global warming and not as a performance that would generate a critical consciousness aware of the instrumentality producing such a reality. To quote Virilio again,

"Ecological catastrophes are only terrifying for civilians. For the military they are but a simulation of chaos, an opportunity to justify an art of warfare which is all the more autonomous as the political State dies out."[58]

The purity of the cold, calculating character of electron optics is related in its essential logic to the transcendent view of an idealized nature that so dominates the normative media apparatus. To invoke the language of Kipling again, both are informed by the "peace" of being "beyond all things"—a form of "exalted contemplation" that sees the totality of existence without hindrance or error. This metaphysical ideal of a peace beyond all things saturates contemporary processes of ecological image making. When threatened by neoliberal development, media depictions typically figure the ecosystem as a passive and ideal realm of purity—almost as if to say that the environment is the most diplomatic of participants in ecological struggles because it is unencumbered by human foibles. The environment is typically photographed from above, from a majestic distance that offers stunning images that invoke the emotive rhetoric of the eighteenth-century prospect poets and the tradition of a pastoral idealism of beauty. In this manner, optic transcendence presents a picture of the ideal environment as one set apart from, and even in direct opposition to, its indigenous populations. We can now relocate this splitting of habitations and their human inhabitants by the imaging apparatus of the society of the spectacle: such representative mandates owe their development to the long and complex history of colonial, military, and technological developments in security. The history of the environment in the U.S. academic context, for instance, was for many years, as Rob Nixon has pointed out, concerned mostly with saving and protecting a presumably untainted wilderness. The institutionalized establishment of a sense of place that trumps migration and hybridity shunned the postcolonial complexities of human political and cultural history in favor of a relation to a timeless and solitary communion with nature.[60] Unlike the earthbound human inhabitants—who are depicted as if shackled to their inevitable displacement, incapable of doing anything more than wait for the inevitable—the ecosystem of the Xingu River paradoxically appears as if not bound by its own earth (see Figure 7). Its exceptional association with a universal ideal of purity grants it the status of a metaphysical grandeur that far exceeds the powerless status of indigenous peoples. This assumed position of purity matches precisely the metapolitical position assumed by privileged representational orders that, in their rhetoric of offering the "plain facts of

Figure 7. An image of the Xingu River in the Brazilian Amazon before the construction of the Belo Monte dam. © Amazon Watch.

the situation" for either security or information purposes, are also understood to operate beyond the play of power relations and partisan interests. Such orders mystify their relations to power and the specific ideological contexts of their conceptual apparatuses by purporting to speak in defense of a Nature characterized as being threatened in terms of maintaining its existence in a naturalized order untouched by human beings and the messiness of politics. In these mediatized figurations of nature during the protests of June 2012, the environment is conveniently imagined and deployed as a pristine entity—a default representative setting that, in and through its explicit matter-of-fact representation, obfuscates its origins in the wilderness rhetoric of colonial history that made nature intelligible as an untouched (and thus always under threat of being violated) scene of serenity.

The violence done to that nature is equally presented from a great height above (see Figure 8). In Figure 8, the trees of the rainforest are almost indistinguishable at first and could be mistaken as matchsticks or lines sliced through drying paint by an artist's chisel. The immensity of the felled tree in the second dwarfs the strangely isolated truck and the

Figure 8. An image of deforestation near the edge of the Amazon, close to the Xingu River. Photograph by Mario Tama/Getty Images News/Getty Images.

long strips of what are presumably (and impossibly) clean and long cuts of timber. On the right, the decay of the top of the tree is already melding into the barren and dusty soil of what had once been a thriving part of the rainforest. On the left, the dead leaves of the tree seem as if on the verge of infecting the trees still attempting to maintain their existence against the onslaught of development. In both sections of the image, the damage done to the rainforest appears as if presented as an immediate affront, not to the local inhabitants, but to the dispassionate and distant gods looking down from on high—the same position held by individuals in the nations of the North and the West watching events of injustice unfold as occasions that frequent foreign geographies. In these images, the ecologies being depicted are not of their own occasion: there is no sense of the geography here as the planet's richest and most complex ecosystem, nor is there any awareness of the indigenous population and the long history of their sustained inhabitancy of the region. Coupled with a different, more critical engagement, they might help to galvanize alternative ecological relations. But as part of the society of the spectacle, their work is a model of shock and awe.

The View from Below

This leads me back to Agamben's original provocation: is it possible, given the ubiquity and force of the metaphysical view from above, to use its own mechanisms against itself so as to realize an alternative "positive possibility" that would unchain politico-ecological struggles from their regulatory spectacularization? It would seem that the metaphysical view carries too much baggage for artists and activists to mount successful alternative figurations. As is well known, the theoretical countermove advocated by critical scholars is to provide figurations and representations "from below" (Zinn's popularized "people's history," Foucault's "genealogies," and Hardt and Negri's "multitude," in addition to countless philosophical discussions on the productive possibilities of grounding thought and political agency not in transcendence but in "immanence"). Such refigurations have functioned as powerful and inventive alternatives to the hegemonic politics of the concrete universal.

However, ground-level figurations of ecological struggles do not automatically mean that we have escaped from the metaphysics of spectacularized events. The majority of images produced by the media in the act of representing the Belo Monte controversy almost exclusively targeted the struggles of people from an on-the-ground standpoint. To anyone watching stories unfold in formerly colonized locations, such images are not unrecognizable. Ground shots centered on the depiction of the old and the young in a state of heightened destitution, showed women washing clothes as part of their daily routines in a river soon to disappear, and revealed a modest private businessman whose shop is slated to fail. In fact, the figurations of the indigenous of the Xingu River repeated, in almost exact fashion, the stereotypical images of marginalized peoples confronting neocolonial authorities—angry groups in acts of street protests, determined villagers in tribal body paint, and forlorn children and old tribal members staring directly into the camera as if shocked and incapable of comprehending the powers at work orchestrating their future as accidental victims of neoliberalism's answer to climate change (see Figures 9 and 10).

Such are the favored images of a representative apparatus seeking to turn the struggle for the commons into the newsworthy spectacle of a non-Western, premodern human race in crisis. Far from merely historically recording their soon-to-be-erased condition, these photos portray

Figure 9. An image taken near the time of the Xingu+23 demonstrations, titled "Displaced Belo Monte Children." Photograph by Mario Tama/Getty Images News/Getty Images.

Figure 10. An image taken near the time of the Xingu+23 demonstrations, titled "Displaced Horse and Carriage." Photograph by Mario Tama/Getty Images News/Getty Images.

the inhabitants of the Xingu River basin as locked into their situation—entirely bound to an unforgiving earth and unable to do more than stare from out of their own destitution. This is not to say, of course, that the representation of peoples soon to be displaced from environments that have been an essential part of their culture for centuries should not find their way into dominant media outlets. These and similar pictures still have the power to motivate alternative political efforts to resist the rapaciousness of neoliberal and antiecological acts of development. Nonetheless, such images are not entirely severed from their affiliations with the normative polity and as such are complicit with the representative logical economy of the spectacle-governed order. As such, the ecocritic is required to trace the historical conditions out of which not only these events themselves arise but the ways in which ecological struggles and ecological habitations become imaginable in the cultural logics of comprehension.

The problematic nature of such image making becomes even more acute when we set such representations alongside the naturalized transcendent representations of a nature purified of all its indigenous inhabitants (Figures 7 and 8). In juxtaposing these images, one realizes that we are being invited to comprehend the environment as a site defined according to the (imperial) ideal of a politics-free realm: a neutral space that rises above the antagonisms of all political interests. Though the Western viewer is invited to sympathize with the plight of a violated people, he or she is also being invited to see these people—no matter what their situation—as ultimately poor substitutes for the perfect world of Nature. For one, they do not have the appropriate technological means for metaphysically representing the "truth" of their situation. For another, their suggested destitution tacitly confirms their lack of empowerment, and therefore authority, in dealing with the ecological occasion of their lives, let alone the immense problem of planetary climate change. In this way, the supposedly benign idealization of Nature over and against its indigenous inhabitants is part and parcel of the production, through the apparatus of security, of an entirely frozen, controlled, and secured environment.

"Pare Belo Monte": Transgressive Metaphysics

The images I have been analyzing were not the only ones that entered the media spectacle apparatus. To contrast these figurations of environmentality, the indigenous people of the Xingu River ecosystem produced their own "spectacle," which the media "shot" from a recognizable height (but paradoxically did not recognize). As part of the customary forms of protests organized by the various tribes, fishermen, and local population of the Rio+23 community (demonstrations and marches in the streets), the indigenous performed two political-aesthetic acts. The staging of these acts was purposeful: they could only be seen and understood properly from a great height above (see Figures 11–13).[61] One of the acts took place in Rio, adjacent to the officially recognized global environmental meeting of Rio+20. On June 19, 2012, close to 1,500 indigenous peoples from Brazil met to form an image on Rio's Flamengo Beach during the Earth Summit II hearings. The performance was organized by the Articulation of Brazilian Indigenous Peoples (ABIP)—Brazil's largest indigenous organization.[62] The critical importance of this group designation lies in its naming of an act of free-representative, linguistic-commons performance in its very title: it names an articulation, the event of a struggle to speak, not an essentialized and targetable identity.

Figure 11. An image made of the indigenous peoples' protest against the Belo Monte dam during the Rio Summit of 2012, titled "Rios Para a Vida" ("Rivers for Life"). Copyright Amazon Watch.

Figure 12. An image depicting the indigenous peoples of the Xingu River in the act of staging a protest against the earthen dam recently constructed as the first stage in the building of the Belo Monte dam. Photograph by Mario Tama/Getty Images News/Getty Images.

Figure 13. The visual performance of the indigenous peoples' protest against the construction of the Belo Monte dam, titled "Pare Belo Monte" ("Stop Belo Monte"). Photograph by Mario Tama/Getty Images News/Getty Images.

The other act took place a few days earlier. Construction of Belo Monte was initiated with the installment of a temporary earthen dam across the Xingu River, which began the process of redirecting the flow of water in the Amazon and the reconstitution of the rainforest ecosystem as a source of energy for mass consumption. On June 15—three days before the "Rios Para a Vida" performance on the beach—members of the ABIP dug a trench through the dam in order to restore the flow of the river. This was also a well-thought performance designed to speak directly to the reproductive apparatus of the metaphysical optic machine. Three days later, ABC ran a photo blog of the event on its "Picture This" website. ABC defines the technological-purifying role of this photo blog clearly: "The ABC New 'Picture This' blog showcases the best in class photography from all over the world as well as photos in the news, photo technology and cutting edge techniques used in photography."[63] The force of the images is thus framed and contained within the domain of technological wonder. The only image of this earlier performance reproduced on the blog was Figure 12—a shot depicting a momentary act of resistance suggesting not an organized event but a partially organized group of people exhibiting an emotional mix of triumph, curiosity, boredom, irreverence, and chaos. But Figure 13—where we see the words "Pare Belo Monte" spelled out, again in the bodies of the indigenous—reveals the well-organized artistic and counterrepresentative intent of the demonstration.[64]

What can we honestly say about these performances and their participation in the "SAGE-like" imaging apparatus that imposes a certain understanding of the environment from its metaphysical standpoint of wisdom? Of course, the insertion of a multifaceted political struggle into a critical theoretical lens can oftentimes result in dubious claims. The effectiveness and specific status of the aesthetics of these performances is a real question. For the people involved in the staging of an event and in its representation and for the critic analyzing the event, there is always the risk of seeing too much and of seeing too little. Too little, and one might overlook the possibility that this struggle may not be offering anything more than the *pathos* handed to us by liberal humanism. Too much, and one risks seeing alternatives that may not be there.

Despite these important reservations, we might gamble some provisional observations and offer some claims for contestation. The performance images of "Rios Para a Vida" and "Pare Belo Monte" constitute, I argue, a collective mobilization of the commons—*but* a mobilization that is in excess of the military-metaphysical logistics of perception we have been analyzing. The images produced by the spatially arranged human bodies function by directly confronting the view from above—the representational order of environmentality. "Rios Para a Vida" depicts a First Nations member reaching toward the sun—a reference meant, according to the indigenous website "Rivers for Life," to express the importance of indigenous knowledge and the indigenous preference for "meeting [their] collective energy needs through renewable resources like solar power."[65] "Pare Belo Monte," in turn, constitutes a literal spatial breakage, in addition to its demand for a breakage of customary representations on the linguistic register of human production. The result, I would argue, is an antispectacularized spectacle—one that is both full of pathos and the ecstasy of organizing against a common enemy and, at the same time, parodic and fully immersed in the knowledge that the planet has come under the sway of a distant imperial armature that watches existence from a transcendent point beyond culpability. These performances, I submit, understand fully the neoliberal reformations of those involved in political struggles on the ground as easily identifiable targets of a supposedly apolitical (but what should perhaps more properly be understood as metapolitical) representational apparatus. Thus the two images of multiple indigenous tribes working together to protest Belo Monte participate with, but also constitute a breakage from, the neocolonial optic machine.

The Part of No Part: Beyond Environmentality

This leads me to the main alternative possibilities explored in this book. Despite the supposedly totalizing nature and power of the discourse of environmentality and its ability to foreclose alternative processes of thought through the mystifying action of the Accidental, its power is not absolute, and the accidents it produces are more than inescapable accidents. The state of environmental degradation that is increasingly defining the potentials for existence does not necessary mean we are condemned, as the security society would have us believe, to the spectacularized, foreclosed future of adaptation. The site of the accident, despite its deep complicity with the custodians of the necessary and the possible, has the potential to offer more than the spectacle of an insecurity that forces us to accept greater measures of security. As the security society mounts more examples and images of climate change in an attempt to develop an insecurity that trumps all previous forms of insecurity, alternative expressions and actions *also* appear, and in an abject relation to the spectacularized. To put it differently, the images produced by the Articulation of Brazilian Indigenous Peoples partake of the metaphysical order of the spectacle. Yet they also speak of a different type of performance, one that, to invoke Rancière, involves the act of naming a wrong by those denied a part in the official proceedings of the reigning political order of the police.

As far as I know, Rancière does not discuss matters of ecocriticism. However, his theoretical work is immeasurably important in the context of current environmental struggles, especially in terms of his conception of politics as the "naming of a wrong" by "the part of no part." According to Rancière, the forms of control normally associated with "politics" (as that term is commonly understood in poststructuralist theory—a negative force that imposes limitations on human agency) depend on the constitution of recognizable and accepted "parts" that predetermine the modes of action an actor can take: political parties, ethnic constituencies, branded identities, corporate structures, underrepresented minorities, and named and counted identities in general (or, to use the customary theoretical terminology, "panoptic cells," "positioned subjects," "docile bodies," "standing-reserves," "metaphysical prospect views," etc.). Rancière reserves the term *policing* for these particular formations of the political.[66] In this sense, the asphyxiating, end-of-history domain of environmentality, governed by the regulatory demands of the security society, is not the realm of politics but the realm of policing.

In opposition, Rancière characterizes the authentic domain of (a more radicalized) politics as involving an agential event wherein those that are not given a voice—those who do not have a part—confront the established problematic terrain by naming the particular failure of the order to recognize their heterogeneous, oppositional voice. Politics is thus an activity that comes into existence in and through the naming of a wrong by those who have no part. The idea of the "no part" should not be too quickly understood as simply "the marginalized" or "the excluded." The part of no part names what Rancière refers to as an "appearance" of a difference that has yet to be fully realized or manifested. A prime example would be what happens when one claims, "I am of the proletariat," as opposed to checking one's class on an official document. The proletariat appears as a potential, but it has yet to actually exist in the order that assigns everyone parts according to a specific "class." Thus, as an appearance, "proletariat" is not a part of the identified and functioning parts of existence. At the same time, this nonidentification in the nonspace of "appearance" names the wrong of enforcing an inequality of power along class divisions. The subject who names himself or herself as part of the proletariat and not part of the lower class opens the potential, in this naming, for becoming a different subject. His or her naming of the wrong is not spoken from the preexistent space of the part assigned to him or her as "marginal"; his or her act of redefinition functions instead as a demand for a different subjectivity.

Politics, as Rancière defines it, thus does not preexist the act: "speech . . . causes politics to exist," a speech that "gauges the gap between speech and the account of it" (26). As such, politics "is primarily the conflict over the existence of a common stage and over the existence and status of those present on it" (26–27). It is an activity that makes appear the injustice inherent in the "distribution of places and roles" (28) and the rules governing the appearance of these places and roles: "Political activity is whatever shifts a body from the place assigned to it. . . . It makes heard a discourse where once there was only noise; it makes understood as discourse what was once only heard as noise" (30). In this sense, politics is not about known objects and places but exclusively about form, the appearance of formations in and through the confrontation of police logic and formations with the form of an egalitarian struggle not previously given formation in the field of the visible—a struggle that makes the field of the visible appear as such, as a set of enclosed spaces that must constantly be policed. The "commons" then becomes the space that makes (the struggle of) equality appear in and through the splitting of what counts as an identifiable community: it is a process of "subjectification against identification" (37). In this sense, subjectification escapes the liberal humanist identification of the indigenous (and similar environmental and ethnic struggles) as victims. The naming of a wrong by those not given access to the formations of experience essentially operates through a claim that we might articulate as follows: I am a subject not because a wrong is done to me (I am a victim); my subjectivity is the act of naming this wrong (I am a political actor). In this sense, the space in which politics appears—a space that has yet to be accounted for—activates the open, free space of the universal, as opposed to the concretized space of a militarized universal.

The indigenous of the Amazon, during the protest of Xingu+23, enacted their resistance not just in the traditional rhetoric of human rights and not just in ecological terms but also with a clear indication of the need to confront a global representative apparatus thoroughly saturated with military might and the Western ontology of transcendental knowledge production. Their efforts created a form of symbolic activism that participated in the reigning aesthetics of the global image-production market but differed radically from the recognizable and media-inundated waving of banners and marching in the streets. They were no longer identifiable victims (as the ground-level shots all suggest); they were forming a new, more active subjectivity by demanding access to the representative space of the universal. These performative events brought into reality the political act of naming a wrong, and their performance, though still participating in the representative order, manifests an appearance of a demand not included in the rationale of that order. Theirs was not a "mere demonstration" but the critique of the militarized machine of visibility that functions by transforming inhabitancies into identifiable targets to be tracked—the field of representation that uses the *logos* as a mechanism for identifying what *counts* as worthy of perception. On June 15 and 19, 2012, the Articulation of Brazilian Indigenous Peoples drew attention to the complicity between representation, military developments in technology, and the neoliberal destruction of the environment—to environmentality.[67] Against the SAGEs of the security system—and the effigies of planet earth as a Medea bent on filicide—these demonstrations make existent and palpable the positive possibility of the common good within the enclosing age of environmentality.

2

Inhabitancy, Custom Law, and the Landless

From Enclosures to Energy Security

The advantages resulting from enclosures are not to be looked upon as merely beneficial to the individual, they are of the most extensive national advantage. The improvements in agriculture, that source of all our power, must be trifling without them.

—Arthur Young, *A Six Months Tour through the North of England* (1770)

Energy security has risen alongside climate change to the top tier of the foreign policy agenda—in much of the United States it outstrips climate change in the priority assigned by the public.

—George Pataki, U.S. Council on Foreign Relations

The founders of our country did not believe that the purpose of government is to serve Allah or the environment, but ensure liberty. We need an energy policy that will do so. The first step required for victory is to disarm the forces that are actively attacking us. . . . As the primary financiers of the global promotion of Islamofascism, the Saudis absolutely must be broken. If that is not done the fire will spread out of control. . . . [T]hese objectives can be achieved by instituting an energy policy that makes the oil experts offered by these enemies expendable. . . . The key word is development. . . . [Biofuels] is the way to achieve . . . these objectives. By making a . . . switch from petroleum to alcohol fuels, we can redirect trillions of dollars that are now going to the terror bankers to finance death, and send it to the world's farmers to create life.

—Robert Zubrin, *Energy Victory: Winning the War on Terror by Breaking Free of Oil*

Allah and the environment—these are heady metaphors for the two phantasmatic archenemies of a sleek and aging post-9/11 American exceptionalism. In the new imperialism of security, however, this fantasy of exceptionalism comes with a twist. An essential hatred for environmental protection remains, cemented in place, while a new demon appears: nature as the new terrorist threat to the anthropocentric constitution of "human freedom" (a coconstitution of rugged American individualism and the ultra-laissez-faire practices of neoliberalism). It's true that this odium for nature as an obstinate force that demands human management owes its conception in part to a long theological and scientific history that extends back to the Adamic creed to subdue the land and wildlife of the earth and the Baconian constitution of modern scientific methods of investigation. The Baconian subjugation of nature was part of a larger constellation of intellectual developments that included, as we saw in the introduction, Galilean and Newtonian transformations in mathematics and physics that recoded nature as anarchic and threatening, a progression that later arose in a different form in the social contract theories of Locke, Rousseau, and Smith. In its early modern American manifestation, this particular civilizing of nature appeared in the religious and political imperatives of the Puritan "errand into the wilderness" and the call to build a "city on the hill" that would serve as the metaphysical basis of the New World (well documented by scholars such as Sacvan Bercovitch, Donald Pease, William V. Spanos, and others).[1] However, the "Medeatization" of the environment inherent in Robert Zubrin's collapsing of all possible constitutions of "the environment" into the security society's discourse on terrorism—what Zubrin refers to as "Islamofascism"—reveals (as we'll see when we look at his argument in detail) a warring attitude that even Adam, Bacon, and the Puritans might have found troublesome (having "dominion over the earth" is not quite the same as interpreting the earth as a fundamental threat to liberty and transforming "undeveloped" nondomestic earth into a diabolic breeding ground for terrorists).[2]

As a security mechanism, Zubrin's proposal—to enclose the Third World in order to curb terrorism—is more closely aligned with the Truman Doctrine of 1947, which made "containment" the official U.S. policy toward the global spread of communism. As technology and information historian Paul N. Edwards has shown, the Truman Doctrine defined the entire history of the Cold War—from the RAND Corporation's impact on nuclear development to the "peace shield" of Ronald Reagan's Strategic

Defense Initiative. The containment policy, still operative throughout the Reagan administration and informing its attempts to keep the Soviet "Evil Empire" at bay, further expanded the metaphysics of SAGE (Semi-Automatic Ground Environment) ideological wisdom. Though Reagan's "Star Wars" speech (1983)—the first introduction of the "peace shield"—was immediately criticized as impossible, its symbolic capital helped strengthen the security society and its militarization of environments, both discursively and materially. Materially, the program ensured continual and growing financial investment, funding defensive research projects in supercomputing and particle physics.[3] In turn, these efforts produced technological advancements that were used in subsequent programs that *globalized* America's ability to secure the planet, such as the worldwide Ballistic-Missile Defense program overseen by today's Missile Defense Agency. Discursively, it helped strengthen the fantasy of security at the public level in America's citizens and allies: through it, Cold War fears of apocalyptic, nuclear destruction were reemphasized and America's global security capabilities affirmed.

Both the Cold War policy of containment and Reagan's Strategic Defense Initiative, Edwards argues, were instrumental in expanding the "technological means to project military force across the globe."[4] Zubrin's proclamation is an extension of these globalizations of security in the form of biofuel technological and agricultural development. His proposal hinges on the deployment of post–Cold War fears of insecurity and is designed to secure the Global South as an immense enclosed farm designed for one purpose: to ensure the continuation of neoliberal development and the technological infrastructure of a militarized state ecosystem. The "key issue in energy independence is not electricity. . . . [T]he key issue . . . concerns the availability of *liquid fuels* to power cars, trucks, trains, ships, and airplanes. These systems are not merely conveniences that have become dear to our way of life. They are the sinews of the economy and the fundamental instruments of military strength. During World War II, when fuel supplies of the Axis nations collapsed, so did their war efforts. A modern war cannot be run, a modern economy cannot be sustained, without liquid fuels."[5] Zubrin's plan reflects, in perversely distilled form, the generalized greening of the security society that I have been referring to as environmentality. It encapsulates R. James Woolsey's call for a green patriotism: reduce your carbon footprint while fighting foreign terrorism in the same stroke. It is of a piece with the green-leaning custodians of the current economy, like *Time* magazine's "heroes of the environment," Ted Nordhaus and Michael Shellenberger, and neoliberal activist for "world flattening" Thomas Friedman— self-named "radical centrists" and "postenvironmentalists" who argue that the agenda of liberal and leftist environmentalism not only is ultimately doomed to fail but also constitutes a threat to economic freedom. Directly opposed to environmental restrictions, people like Friedman, Nordhaus, and Shellenberger emphasize the expansion of economic growth, technological innovation, and national ascendency in the face of "fundamentalist threats to American democracy." They thereby manifest their enclosure history inheritance, upholding the great eighteenth-century enclosure campaigner Arthur Young's view of environmental development as an essential component for growth, improvement, and national supremacy.[6] In this new-and-improved discourse of twenty-first-century "energy security," one can become green while still dismissing the environmentalist demands for regulation and transformation that would seek to bring about a more sustainable relationship to the planet.

As I argued in the introduction and chapter 1, the struggle between ecological preservation and the unrelenting greed of neoliberal development makes the Brazilian Amazon an especially troubled location. Though my focus will be on Brazil in this chapter, the discourse of environmentality cannot be fully appreciated without tracing its movements across different geographical locations and through different historical moments of anthropocentric development. The contestation over anthropocentric energy needs—the future plans to construct dozens of dams in the Amazon, for instance—and the fight to maintain the indigenous spaces and the ecosystem of the rain forest generate paradoxical visualizations that bring out some of the tensions inherent in the globalized discourse of environmentality. These tensions also manifest themselves in a much longer and more complex history of political–environmental struggle that extends beyond the context of Brazil's landscape to the context of America (in that America has inherited the Portuguese and English enclosing ecological influence on Brazil). Because of its extensive colonial history of agricultural development (the colonial plantation system and the sugarcane industry), the landscape of Brazil serves to highlight the specific globalized nature of the discourse of enclosure—its relation to slavery and how that discourse informs today's environmentality. As I argued in the introduction, the discourse of enclosure is far from being a curiosity of the past. It has generated a platform of existence that continues to have a significant effect on our current economic, political, and ecological relations. In this chapter, I will take a brief look at the history of the enclosure movement, its material and discursive development, and its direct relation to the rise of the environmentality of "energy security." I will conclude by examining the alternative possibilities offered in the counterhegemony of the Brazilian Landless Workers Movement, commonly referred to as the MST (Movimento dos Trabalhadores Rurais Sem Terra). These alternative possibilities I ground theoretically by considering the historical importance of one of England's most important and most overlooked legal events: Gateward's Case of 1608, which inaugurated a war against inhabitancy that continues to this day.

A Brief History of the Enclosure Movement

The economic and military tactics previously listed stem from a paradigmatic pattern of thought that extends to an ecodevelopmental past not often recognized, and the immense impact of this past on the present is hardly comprehended. I have been identifying that past in terms of the seventeenth- and eighteenth-century accession of the discourse of enclosure. Many historians of enclosure, though they understand the movement to have widespread, planetary implications, emphasize its economic character in their analyses: the onset of private property, individualism, entrepreneurship, and urbanization as part of an organizational structure opposing the old noncapitalist, subsistence agriculture. E. P. Thompson, for instance, summed up the movement quite aptly by characterizing it as "class robbery."[7] This economic aspect of enclosure is, of course, an essential component of its ontological makeup, but the enclosure movement is equally important for its impact on the history of colonial and postcolonial concerns for land and environmental relations; for the constitution of human subjectivity; for the rise, in the twentieth century, of a certain epistemology of agricultural development; and for the development of a certain pattern of thought that empowers the paradigmatic shift identified by Foucault—from disciplinary societies to the globalization of the security society.

The historical specifics of the movement from which these connections can be drawn are often elided when environmental activists, advocates, and scholars mention the "enclosure of the commons."[8] Many characterize the movement as the invention of an order entirely opposed to the old feudal system, in turn lumping the unenclosed agricultural architecture of strip-field farming, the commons, and the open-field system into that same feudal paradigm. The commons system is thus equated with an organizational structure known for its opposition to modern ideas of liberty based on the conception that feudalism offered only two possibilities for human existence: a national subject, or countryman, was either a privileged lord of a manor or a serf chained to the land. This highly reductive characterization of the commons system was further compounded by the mythology of the commons generated by the American ecologist Garrett Hardin in his essay "The Tragedy of the Commons," published in *Science* in 1968.[9] Hardin argued (1) that the commons were a less-advanced form of social existence, one that existed without rules or regulations; (2) that, without regulation, humans would breed excessively, resulting in a Malthusian catastrophe; and (3) that because humans were fundamentally self-interested and at war with one another (and here Hardin discloses his complicity with proenclosure advocates Locke and Smith), this unregulated social space of existence would result in the overexploitation and ultimate destruction of natural resources. Nothing could be further from the truth.

Contrary to received opinion, the communally organized and highly regulated social system of the commons predated the feudal system by a number of centuries. As we will see, this means that Hardin's single-minded argument stems from the categorical error of imposing postcommons humanity (i.e., a humanity accustomed to entrepreneurial acts of enclosure for capital accumulation) on a paradigm of production that has nothing to do with an enclosure-oriented entrepreneurship. To put this differently, the humans he sees ruining the commons are the humans constituted according to a form of subjectivity that only became the norm *after* the discourse of enclosure came to dominate social existence. It is now known that common rights, often called "use rights" (the liberty to have access to land that was considered open to everyone for grazing, gathering fuel and supplies, erecting cottages, etc.), stem from a much older complex system of a collective cultivation of land that was prevalent not only in England but throughout Europe and elsewhere. Under this system—often referred to as the "Saxon system"—there was no sense of individuals possessing the land as we understand the word *possession* today: as highly guarded, private property in need of security. The subject having access to the land was not an entrepreneur bent on accumulating capital; the subject was considered to be an "inhabitant" of the land and accessed the land's unrestricted openness from a standpoint of sustenance rather than as an investment in property development. This difference in the essence of subjectivity—based on inhabitancy rather than entrepreneurship—is the key to understanding the problem with Hardin's assessment of the commons system. Hardin assumes the inherent essence of the human subject to be capitalist and consequently that any open-field system with access to all will result in the exploitation of environments. His "rational" argument lacks a deep historical awareness of preenclosure sociopolitical reality. He overlooks the three-century-long war against inhabitancy that ended with the new specifics of the individual entrepreneur gaining supremacy over previous forms of subjectivity. We will look at this war, which was waged on the register of the judicial system, in some detail in this chapter. Hardin, as we will see, bases his scientific results on a form of human subjectivity specifically manufactured for the discourse of enclosure, a form of subjectivity that makes possible the various forms of capitalist entrepreneurship that follow.

In opposition to the generalized modern (proenclosure) view of the old commons system, the use rights offered through the concept of what was called "land tenure" was an inherent part of the Saxon, not the feudal, system. The "openness" of land tenure is often recast from the post-Saxon stance as offering a less-advanced economy of social relations; it is characterized as lacking the mechanisms necessary to "improve" the environment and regulate its (energy) resources. But both "improvement" and the regulation of resources are imperatives that arise with the creation of postcommons enclosure rationality. Land tenure, as scholars increasingly point out (W. E. Tate, Gordon Batho, and later Jeremy Rifkin and Elinor Ostrom), involves complex mechanisms of preservation specifically designed to guard against the entrepreneurial exploitation of resources. In many open-field communities, for instance, villages employed intricate mechanisms of seasonal distribution and redistribution (such as the "mead stick" system) that were designed to ensure that no single person would gain a monopoly over the land.[10] These mechanisms were designed to ward off the unequal accumulation of power and the exploitation of resources. As use rights of the Saxon system began to erode during the feudal era, tenure began to disappear, and two social groups of farmers began to form: the first were "villeins" of the manor, who held certain rights to the land and to their cottages but could be evicted by the lord, and "serfs," who had no land holdings and worked as "bondsmen" of the lord but who still had access to common rights.[11] From these two categories, other subcategories began to form, linking inhabitants of the land to the ruling arm of the feudal lord. Contrary to Hardin's assessment, history has shown us that it is enclosures—the dominant paradigm of modernity—that contribute to the exploitation of resources and the overpopulation of the planet. The very idea of a cash crop—an environmental "improvement" that compromises biodiversity in favor of anthropological gain—depends on the logic of enclosure. As many historians, ecologists, economists, and political theorists—such as Nobel Prize winner Elinor Ostrom—have pointed out, the commons system is far from being a self-evident "tragedy."[12]

Awareness of the effect of enclosures in the present is often undertheorized due to a general annexation of the phenomenon to the past (though recent efforts have begun to discuss its impact on the present).[13] Some consider the movement to have been a phenomenon of Thomas More's era; others reference it as being primarily an event of the eighteenth century. Official historical record, however, places the last British Parliamentary Act of Enclosure in 1914.[14] But the enclosure of the commons did not end there, even within the confines of England. In 1947, King George VI empowered the Ministry of Agriculture to enclose any common requisitioned during World War II and shown to be capable of substantial agricultural productivity.[15] Even later than this, a 1967 Greater London Council (General Powers) Bill in Parliament contained a clause that gave Metropolitan Water the right to enclose and extinguish common rights in an area in Epping Forest. And in 1962, the lingering mandates of a two-century-old enclosure from 1795 brought about the eviction of a Somersetshire cottager—on the grounds that he was squatting without any legal rights.[16] As England expanded its empire beyond its immediate borders, the enclosure movement was transplanted globally—to India, America, South America, several African territories, and elsewhere.[17] As the privatization of land and a cash-crop economy become the key agricultural features of colonial expansion, other European countries began to deploy these methods in the expansion of their empires overseas. Today the neoliberal enclosure of the commons continues with unrelenting force, including large-scale international agreements such as the structural adjustment policies of the World Bank that destroyed communal land relations in nations across Africa;[18] the repeal of the ejidos, or communal lands, in Mexico as a result of the free trade imperatives of North American Free Trade Agreement; and the privatization of land, space, and resources in more than eighty countries by the International Monetary Fund and World Bank since the Washington Consensus, to name only a few.[19] Portugal's colonial enclosure of Brazilian land and the erasure of complex ecosystems in favor of large plantations of cash crops such as sugarcane stand as essential historical ironies that are now producing present-day environmental conflicts (sugarcane, it should be remembered, is the key rationale for the "greening" of Brazil's formerly oil-dependent culture).

In a similar fashion to the way in which Foucault discusses the transformation of sexuality into discourse during the Victorian era and early twentieth century, enclosure as a historical phenomenon becomes a major epistemological concern of knowledge production in the modern university. Enclosures and the commons, as two opposing forces, became a significant object of knowledge. Innumerable volumes and extended historical analyses were produced, such as the magisterial, multivolume *Agrarian History of England and Wales*—a work of dozens of scholars from leading universities that was the outcome of an official agricultural "Advisory Committee" formed in 1956.[20] This epistemological constitution revealed, perhaps for the first time, the extent to which the enclosure movement brought about the securitizing development of what we might call "panoptic land"—that is, the installment of a general economy of land administered by a centralizing agricultural and political ideal. As many agricultural historians have argued (Mingay, Thirsk, Turner), the state's knowledge and control of land prior to enclosures were, despite various efforts, limited. The mapping of land, as we saw in the introduction, first began on a wide scale because of acts of enclosure. But more than that, the enclosure movement inaugurated an abstracted mechanism of government that could be applied to any village. Here it is important to remember the importance of Foucault's emphasis, in his critique of Bentham's invention, on the naturalization of the panopticon as an ideal mechanism of management and regulation, one that differs from the overt appearance of a momentary application of power:

The Panopticon . . . must be understood as a generalizable model of functioning; a way of defining power relations in terms of the everyday life of men. . . . [It] is the diagram of a mechanism of power reduced to its ideal form; its functioning, abstracted from any obstacle, resistance or friction, must be represented as a pure architectural and optical system: it is in fact a figure of political technology that may and must be detached from any specific uses. . . .

In each of its applications, it makes it possible to perfect the exercise of power. It does this in several ways: because it can reduce the number of those who exercise it, while increasing the number of those on whom it is exercised. . . . Because, in these conditions, its strength is that it never intervenes, it is exercised spontaneously and without noise, it constitutes a mechanism whose effects follow from one another. . . . The panoptic schema makes any apparatus of power more intense: it assures its economy (in material, personnel, in time); it assures its efficacity by its preventative character, its continuous functioning and its automatic mechanisms. It is a way of obtaining from power "in hitherto unexampled quantity", "a great new instrument of government". . . .

It can be integrated into any function (education, medical treatment, production, punishment); . . . it can constitute a mixed mechanism in which relations of power (and of knowledge) may be precisely adjusted, in the smallest detail, to the processes that are to be supervised; it can establish a direct proportion between "surplus power" and "surplus production." In short, it arranges things in such a way that the exercise of power is not added on from the outside, like a rigid, heavy constraint, to the functions it invests, but is so subtly present in them as to increase their efficiency by itself increasing its own points of contact.[21]

As Foucault details here, the panopticon was an apparatus designed to steer subjects toward a normalized optimum of behavior, thought, and performance. It was a process of making subjects into self-disciplined selves in and through the internalization of an ideal; it was a form of power not "added on" but made possible in and through the constitution of subjects and entities that were themselves part of the mechanism of the ideal. Although Foucault certainly did not have environmental concerns or land relations in mind in his theorization of Bentham's model for social government and behavioral control, we can nonetheless find powerful parallels in the enclosure movement's demand for new architectural structures of increased production, optic surveillance, and the erasure of "inhabitancy" in favor of the new sovereign subjectivity of the enclosing landlord.

In the case of enclosures, the entrepreneur is the subject constituted as the essential human half of a mechanism that transforms ecosystems into a "surplus" of power and production. In the case of the land itself, the ideal was the improved enclosed land working for the demands of gross national production, state intelligence on its agricultural holdings and citizenry, and military data on national geography—in other words, a generalized panoptic apparatus of governmentality attached indissolubly to the land. The individual that cultivated that land stood out as the economically useful and politically obedient subject—the modernized subjectivity that was characterized as a rational being. This new modern subject was directly placed in opposition to those living in unenclosed spaces (irrational vagabonds, gypsies, "masterless men").[22] In turn, the land comes under the sway of new methods of knowledge-power production: greater attention to geographical details, close analysis of soil, establishment of clearly defined borders, insertion of crops into a new demand for greater surplus, and so on. Overall, the land is inserted into a system of government demanding a nationalized increase in efficiency tightly linked to a more expansive connective state apparatus—a closed-loop system, as we will see, that ultimately produces its own accidents and, in turn, the "empirical need" for enforcing widespread measures of adaptation.

Panoptic land—like the creation of disciplined, "docile bodies" in the eighteenth century analyzed by Foucault—is land broken down and reterritorialized for purposes of being disciplined according to a system of optimized efficiency: terms such as *cultivation* and *improvement*, for instance, are part of a lexicon deployed ad nauseam throughout the seventeenth, eighteenth, and nineteenth centuries. In addition to emphasizing the constitution of a self-disciplined subjectivity, Foucault characterized panoptic mechanisms as systems of ordering that relied on increased visibility, in direct opposition to the old punitive systems, which imprisoned people in dark spaces such as dungeons. Panoptic mechanisms of control and calculation involved widespread systems and instruments of surveillance: land was lit up, so to speak, by the massive amounts of mapping that an enclosure act required. The age of parliamentary enclosure, in other words, discovered land and the environment in general as objects and target of power. During this century, the environment became subject to an overseeing panoptic principle of efficiency. The demand for a "general view" of the agricultural landscape of England (the title of many agricultural volumes published in the eighteenth century) reflects the mandate to increase the productive force of the land (its "energy") in economic terms of utility. At the same time, this disciplinarity implements heavy restraints on the land by diminishing "errant" forces (less productive common land and "wasteful" itinerant movement between fields under the former strip farming methodology) to maintain political obedience. Panoptic land enforces a unidirectional pattern of thought, behavior, and development by coupling an "increased aptitude" with an "increased domination."[23]

If we consider that enclosures began before the development of capitalism during the transformation from the Saxon system to the more militarized manorial system—made into law in 1237 by the Statute of Merton—then enclosures arise as a phenomenon within the context of a certain militarized outlook.[24] The great manorial lords desired the legal right to enclose for the purpose of increasing their wealth and, by extension, the ability to direct resources toward their defensive capacity, in addition to the pursuit of privileged forms of entertainment (such as the hunting of game normally associated with these early manorial enclosures). The rise of capitalism that followed feudalism continued this erasure of the communal essence of freeholders, copyholders, leaseholders, and so on by creating two categories of individuals: those that had the power to enclose and those that were dispossessed by acts of enclosures. The enclosure movement thus created a disposable or subaltern (to use the postcolonial term) class of humans in the process of constituting a new, political form of the human based on the anti-inhabitant individual whose rationale for existence was the economic production of a "greater yield" (the key agricultural term of the discourse of enclosure).

To put this in terms of the security society logic that we have been analyzing, enclosures created what Agamben has identified as "bare life" (*homo sacer*) in the process of constituting a new form of subjectivity endowed with a "qualified [political] life" (*bios*).[25] Resistance to enclosures by this new dispossessed class of inhabitants reduced to bare life was widespread throughout the nation and continued for several centuries: Kett's Rebellion in 1549, the Midland Revolt and the Newton Rebellion in 1607 and the rise of the Levellers, the Western Rising of 1626–32, the beginning of the Digger movement in 1649, the Swing Riots of 1830, the establishment of the London Corresponding Society in 1792 (advocating for parliamentary and land reform), the Otmoor Riots of 1830, the Dean Forest Riots of 1831, and others.[26] Parliament's reaction to these enclosure riots was to pass a series of legislative acts designed to prevent villagers and farmers from forming in groups of more than three.[27] By 1590, popular protests of enclosures in the form of hedge leveling had created such widespread fear that the charge of "revolt" in these protests was raised to the level of "treason against the King."[28] The post-Restoration government began to openly sanction acts of enclosure, revisioning them as a way to cultivate the land more productively.[29] In the mid-seventeenth century, the last antienclosure bill was presented in the House of Commons and rejected.[30] The form of these protests would eventually evolve during the Industrial Revolution, as enclosure became connected to technological improvements in agricultural machinery. Enclosure created the wealthy class of tenant farmers who favored such human-replacing inventions as the threshing machine. This led to the Swing Riots, which began in the summer of 1830 and by December had spread across the whole of southern England and East Anglia.[31] Continual resistance throughout the rise of the new enclosing economy required the state to pass a series of antirioting measures. As the privatization and "improvement" of agricultural ecosystems became the accepted paradigm of existence, the state would call upon its police and military forces to protect the enclosure of the land, thereby turning more in the direction of establishing a security society.

By the late eighteenth century, the movement was hailed by the growing entrepreneurial class as one of the greatest advances in land development, regarded increasingly as a revolutionary scientific development and generally thought to be more efficient and cost-productive—increasing the production of grain yields on the farms of some of England's counties from ratios of three to one to twenty to one.[32] An efficiency-driven relationship to the land had now taken hold of the general public consciousness. An act of enclosing a portion of the land guaranteed that formerly unconnected communities would thereafter be tied to a growing gridwork of commercial enterprise. The isolated, nucleated local village became part of an emerging international market.[33] The land was being reterritorialized into the new "world picture" of the nation, which favored development over sustainability and the establishment of strong and expanding ties to a growing international economy. This developmental, market-based paradigm was coupled to the military concern for security: the eradication, for instance, of the lowland "wastelands" occurred during the wars with France, and the conquest of these wastes became synonymous with the conquest of France itself. In 1803, Sir John Sinclair wrote, "We have begun another campaign against the foreign enemies of the country. . . . Why would we not attempt a campaign against our domestic foe, I mean the hitherto unconquered sterility of so large a proportion of the surface of the Kingdom? . . . Let us not be satisfied with the liberation of Egypt, or the subjugation of Malta, but let us subdue Finchley Common; let us conquer Hounslow Heath; let us compel Epping Forest to submit to the yoke of improvement."[34] The control of "waste" at home reflected the desire for national domination of land abroad, which, during the nineteenth and twentieth centuries, came to be associated with the control of colonies around the planet.[35] Developing into a widespread discourse and a new ontological ground plan for human existence in the eighteenth and nineteenth centuries, land began to be represented universally as a raw material that needed to be overseen by the nation-state and protected by its military—an ecopower that followed alongside the constitution of humans in terms of biopower. By 1914, the British political machine had extended the enclosing land apparatus in one form or another to some 55 percent of the globe in the form of colonies, commonwealths, dominions, and protectorates. The same history of resistance to enclosure within the British Isles was repeated, with little farce and most often with great tragedy, across this geography of colonial exploitation.

As I will discuss in more detail in subsequent chapters, it was during this time in the twentieth century that both the formal and the natural sciences of agriculture, geography, economics, and engineering came to be intimately coupled with the military machine, creating the particular architecture that governs today's environmentality. As we saw in chapter 1, the nineteenth-century imperial enterprise to map England and India was an essential component of the military machine. Such efforts were made possible by previous developments in surveying technology brought into existence by the enclosure movement's new epistemological demand for a mathematically precise geographical and topographical science. Enclosing a plot of land involved extensive surveys and lengthy disputes of pathway rights, boundaries, and access. Each enclosure "act"—the legal document finally produced after months of negotiation—required a series of surveyors' maps. Alongside these maps, numerous surveys of the land were published, such as Edward Hasted's twelve-volume *The History and Topographical Survey of the County of Kent*, compiled between the 1750s and 1790s. Hasted's survey, like others of its kind, identified areas of the landscape according to licit and illicit spaces: enclosed spaces were characterized as "remarkable, beautiful, and pleasant," "lively," and full of "grace and gaiety."[36] Open areas were labeled as "promiscuous" and the inhabitants of open areas as "wild, and in as rough a state as the country they dwell in."[37] Ecosystems came to be identified as "useful" or "bare." Military-oriented parliamentarians not only applauded the enclosure movement as an agricultural advancement in terms of its ability to produce "higher yields"; they also praised the physicality of enclosures (hedges and stone walls) as a defense against foreign encroachment on domestic and colonized soil in addition to potential threats from the "homeland." These surveys, often titled "General Views," reflected the development of a new economy and policing of land. Set within the context of an expanding colonial empire that saw the environments of distant colonies as threatening and in need of improvement, the enclosure of unimproved land at home became a matter of national security.

By the eighteenth century, England began to lead the world in grain yields, producing, in general, harvests of 50 percent above any other European country.[38] The 1797 edition of the *Encyclopaedia Britannica* boasted that "Britain alone exceeds all modern nations in husbandry."[39] The scientific justification of increased yields that followed acts of enclosure led to larger forms of enclosure called "engrossment," which saw the amalgamation into large farms of formerly small peasant holdings. As a result, small farmers suddenly found themselves transformed into a landless class of wage laborers subject to greater forms of economic and political insecurity. Wage laborers, unlike their village-laborer predecessors, were subject to market fluctuations—experiencing a drop of 50 percent in real wages between the years 1500 and 1640. And unlike village laborers, wage laborers could not rely on the produce from common land as an alternative means of support.[40] These specifics, buried within the long-forgotten historical occasion of the movement, reveal that enclosure has less to do with the search for methods of "scientific advancement" and "historical progression"—even though these are the terms ceaselessly invoked in proenclosure historical analyses of the enclosure movement—than with the installment of a new oligarchical strategy of social control and political domination. This forgotten phenomenon, which was also commonly referred to as the "disappearance of the small farmer," was at its most concentrated at the height of England's imperial order in the Victorian era.[41] Throughout the nineteenth century and in the early twentieth century, these meticulous strategies of control and mastery of land at home were transplanted abroad, thus enabling the discourse of enclosure to expand to a global and colonial dimension.

The power of the enclosure movement is especially effective at the levels of epistemology and historical representation. The demand for documentation was an inherent part of the enclosing apparatus, which required unprecedented levels of information to be gathered. This information in turn acquired a new level of significance. This demand for documentation formulated an entire field of knowledge subsequently made available to state authorities and to the scholar, knowledge that would have previously been meaningless and even nonexistent. The movement manufactured the modern necessity for "information gathering" (enabling bureaucratic administration and control) on a grand scale: an official "enclosure award" (the term for the actual legal document) for even the smallest farmland was typically upward of fifty broad sheets and contained detailed information compiled by the state's enclosure land surveyors. As the enclosure historian William Tate points out, this was the first time in English history that a written record mapped out the precise dimensions of highways, villages, and individual farms. More than this, the awards also contain information about specific numbers of population, types of territorial disputes, patterns of urban and rural migration, locations of useable and "waste" land, acreages of forests, differentiations in soil, and censuses of citizenry for purposes of taxation. In many cases, the enclosure award served as the founding charter for villages that for centuries had no official form of documentation.

Thus, in addition to transforming the human–land nexus from a sustainable relation[42] to one of exploitation for the improvement of resources, the enclosure movement generated a usable and efficient mechanism for a globalized form of representation based on bureaucratic ideas of information and documentation that we now consider self-evident. This in turn made "documented" land and consequently documented citizens somehow more civilized and meaningful than land and peoples in geographical locations that had yet to enter into the enclosure registry. Globalized documentation easily went hand in hand with the extension of imperial structures and with the attitudes throughout Europe and America that informed colonial expansion. Annexed to the periphery, away from the center of "proper" culture, the undocumented commons—and the people living on them—were given an inferior status that was seamlessly coupled to the "uncivilized" and "wild" imperial colonies of Africa, India, South America, and the East Indies. Enclosed spaces, on the other hand, were coded as "cultivated land"—a usable commodity for the new age of environmentality.

It is in this sense that the early environmentality manifested in the discourse of enclosure should be comprehended as a form of what Foucault identifies as "governmentality": the essential imperative to "improve" the land and the coupling of substantive forms of documentation create a new, more modern nation-state desire for the productive management of individuals, populations, and national space. Although Foucault began to use the concept of governmentality late in his career and the full lecture from the course that develops the term at length was not published until well after his death,[43] many have since extended the concept and expanded its usefulness for comprehending the specific ontological character of the governing of populations that arises with greater emphasis over the course of the last two centuries.[44] Governmentality names a system that places emphasis on productivity and usefulness as opposed to disciplinary mechanisms of repression and overt control made possible in and through the extension of rules and regulations. According to Jacques Donzelot, urban planning advisor for the Ministry of Land Management in Paris and founder, along with Foucault, of the Prison Information Group, governmentality involves an effort to actively shift, for instance, the concept of work in the mind of the worker: it becomes necessary to move work away from an awareness of it as a laborious activity toward being part of a meaningful and rewarding enterprise.[45] Transforming work into a positive activity enables at the same time the internalization of the act of working as an act of personal self-development. Advocates of enclosure employed similar tactics when they compared enclosed spaces, and the people that worked them, to a meaningful national enterprise. The measurement of a successful improvement of land was also the measurement of the productive capacity of the individual and a sign of the individual's patriotic support of national productivity.

This extensive history of the enclosure and privatization of land, however, amounts to no more than a preamble to the kind of massive planet-wide enclosures we are seeing today. Throughout the nineteenth and twentieth centuries, the governments of Brazil, Argentina, Columbia, India, the United States, Canada, Mexico, South Africa, Kenya, and elsewhere would begin to adopt the policies of land enclosure. In many instances, this would lead to the almost total eradication of indigenous peoples in these regions and to extensive amounts of ecological transformation that demanded greater, accidental forms of adaptation.[46]

The Twenty-First Century: The True Age of Enclosure

We can best understand the history that led to environmentality by first looking at Brazil's historical relation to the globalized security society, its connections to American interests in security and neoliberal developments, and in general the manner in which the country must contend with the planetary discourse of environmentality. Brazil's ranking in terms of global economies is increasing with each new multinational coupling, yet its economic inequalities continue to expand. The wealthiest 10 percent of the population possess 45 percent of the nation's income; the poorest 20 percent possess less than 3 percent.[47] Despite the now more than thirty-year agrarian reform efforts of the landless, Brazil continues to have one of the world's highest examples of land concentration: "According to Brazil's land registry, 1.6 percent of the landholders control 47 percent of the nation's farmland, whereas a third of the farmers hold 1.6 percent of this area."[48] More than half the population lives below the poverty line. In recent years, 14 percent of the population has suffered from hunger, and estimates put the number of landless anywhere between 3.3 and 6.1 million families. The state continues to support large landowners at the expense of the small. Despite struggles for agrarian reform in the judicial system, liberalized trade policies continue to favor partnerships with agro-food conglomerates that control the global market. Political influence lies in the hands of the tripartite structure of the nation's large landholders, the state, and global corporations.

The political representation of the landless, small farmers, and the elite is based on the number of federal deputies allocated to each, and the representation of the elite landlords far outweighs that of the poor: "2,587 greater than that of landless peasants."[49] Despite frequent negative reports of the MST in the popular press, the organization gained widespread public support in the mid- to late 1990s, after two police massacres of landless in the Amazon and a sixty-four-day march of 1,300 landless to the capital, which culminated in the additional attendance of close to 100,000 supporters. However, in the early years of the twenty-first century, the World Bank, working with the Cardoso government, introduced new market-based measures to land distribution. These actions damaged the movement and its achievements, perhaps more severely than before. Such policies were even active under the Lula government, which also pushed for greater land distribution in the Amazon.

The MST arose out of the impossible living conditions brought on by the policies of three decades of dictatorship, by the neoliberal exploitation of the land's resources, and by the continuing colonial legacy of a low-wage laboring system formerly set in place by Spanish- and British-influenced Portuguese structures of colonial development. The movement began with a number of farmers who, having been pushed to extreme impoverishment, decided to set up camp and become squatters on the edge of agribusiness farmland. At this early stage, the movement had a difficult time contending with aggressive representations of its existence as an anarchic force, which continue in the present. The conservative media in Brazil often brand the MST as a "criminal mob" and characterize its educational institutions and tactics as mindless propaganda and indoctrination.[50] Brazil's most popular and conservative magazine *Veja* has a long history of demonizing the MST. In 2004, in the wake of 9/11 and manufactured fears of encroachment by Muslim culture throughout the United States and Latin America, it printed that the MST educational system was identical to a "radical Muslim" boarding school—that it was teaching its children hatred and forcing them to view all non-MST people as "infidels."[51] The post-9/11 environment made it easy for the popular state media to characterize the organization as a terrorist group and a threat to liberal democracy. In addition, recent evidence reveals that the MST's movements are being heavily monitored by the Brazilian military.[52] As Brazil's ties to the globality of neoliberalism grow stronger, the MST must also confront increasing characterizations of its efforts as signs of an "autocratic, violent, shady revolutionary organization" that poses a threat not only to the Brazilian state and democracy but also to American national security.[53]

The landless workers movement is steeped in a long history of Brazilian political, military, and economic control. In the early 1960s, after many years of political dictatorship, it appeared as if a more egalitarian system would come into existence. However, the landless and the poor were not well organized. A number of peasant and left-leaning union organizations existed, but their effectiveness was limited. These organizations, unlike the MST, did not develop their own language of resistance but used the language made available to them by the Brazilian Communist Party (the party itself was of little help to the unions and peasant coalitions). Despite the limited potential for the poor to gain some momentum, the powers of the North, mainly the United States, began to take an interest in their regional activities.

In the midst of the Cold War climate, the United States was apprehensive about the potential for movements of poor people to develop and grow in strength. It felt that such organizing of the poor would only contribute to the "domino effect" of spreading global communism. Both the Johnson and the Nixon administrations openly supported Brazil's military dictatorship. In the days leading up to the military coup, President Johnson went so far as to say to Undersecretary of State George Ball that the United States "should take every step that it can" to support the takeover of the democratically elected government.[54] The United States considered military regimes to be an ideal model for Latin American governments, and the Cold War consensus was that Brazil would become a "Latin-American China" if action was not taken to stem the tide of activist efforts. MST scholars Angus Wright and Wendy Wolford detail the changes in the organization of the social structure in the wake of the coup: "After the military coup, organizations in the rural areas came to a virtual standstill. The trade unions that had started up in the 1950s and 1960s were violently repressed, and also infiltrated by governmental workers. The Peasant Leagues were broken up, their members intimidated and their leaders exiled."[55]

The elite of Brazil also had reasons for putting an end to any potential for a poor uprising. Working with the U.S. military, Brazilian landowners, businessmen, and military personnel joined forces to overthrow President João Goulart. After the coup, the Brazilian military began to systematically attack peasant groups, trade unions, and student organizations. Despite the violent methods the military used—including torture and execution—organizations of poor farmers continued to protest, until the military decreed "Institutional Act No. 5," a law that made any form of criticism or opposition to the government illegal. The next two and a half decades would see the military dictatorship working more and more closely with global economic organizations, eventually producing the so-called Brazilian miracle (becoming a world economic leader) so praised by neoliberal conservatives in the United States today. Many scholars have criticized the history of various U.S. administrations and their connections to forms of military dictatorships. Though it is not acknowledged in these terms, military dictatorships and their inherent obsession with security, strict supervision, domination, and bureaucratic control offer distant neocolonial apparatuses the certainty that the population will not change its economic structure. They serve as deployable "force elements" (to use the military metaphor) that serve to support the neocolonial interests of Western nations (in chapter 4, I will explore this structural maneuver at length in the terms of the military strategy of "force elements," also known as "directed telescopes," developed by the United States in the nineteenth and twentieth centuries). These force elements of military regimes may not directly owe their allegiance to a more powerful nation, but the fact that they are a military polity that oversees the governing of the nation ensures that all operations—political, economic, social, and so on—are under the control of the war machine and within the limits of the security society paradigm. The dictatorship thus functions as an extended armature of the neoliberal security state, operating like a staff of soldiers sent from the command center to various locations on the field of battle. Within this economic-military ontology, the primary goal is to secure the territory of the ecosystem for raw material extraction—an objective directly opposed to the concept of inhabitancy.

As mentioned in the introduction, one key location within the borders of Brazil that has had one of the worst histories of destructive development projects and the worst record in terms of environmental degradation and quality of life is the country's northern region, including sections of the Amazon. The north of Brazil is steeped in four hundred years of colonial rule and development. Even to this day, the sugarcane farms begun by the Portuguese colonists dominate many northern habitats. These colonial relations to the environment have expanded exponentially in the neocolonial age of environmentality. Today northern Brazil is home to Camaçari, the Western hemisphere's largest petrochemical complex. Camaçari comprises more than fifty companies, and its fifty thousand employees work in the proximity of chemicals, such as benzene and alcohols, that affect the Amazon's central and peripheral nervous system. Workers operate with little awareness of these chemicals' toxicity.[56] The complex is owned by the Brazilian multinational company Brasken, which sells its chemicals to such companies as Dow and Innova. The multinational was formed in 2002 and since then has acquired several other Brazilian companies through a series of mergers. In 2010, it acquired the U.S. chemical company Sunoco.[57] After the completion of its "Green Ethylene" plant in September 2010, Brasken was hailed as a leader in sustainable chemical production. (Ethylene is a polymer produced in various forms of plastic used for food packaging, home appliances, cleaning products, toys, etc.) The "sustainability" of this chemical means big business for the company, and it is already in the process of acquiring several transnational contracts.

This relation to the Amazon stems back almost half a century. Former president Admiral Garrástazu Médici, the nation's leader during the military dictatorship, attempted to divide and conquer the landless of Brazil by targeting the Amazon as an ideal space for colonization by peasants. In 1970, he proclaimed the forest to be "a land without people for a people without land." Médici's vision was to solve the country's peasant problem by sending millions of landless to live in the Amazon forest and constructing, in the process, the Transamazonian highway.[58] (The government's solution of relocating the landless to the Amazon continues today.) For many, the Amazon is attractive primarily for its timber. Greenpeace estimates that between 60 and 80 percent of all logging operations in the region are illegal.[59] The colonization of the forest threatens the biodiversity (the world's richest) of the Amazon, and the loss of trees releases CO_2 into the atmosphere. Not as commonly known, however, is the relationship between the Amazon and the production of freshwater. According to Wright and Wolford,

> The Amazon river carries more than 20 percent of all the river water that flows to the sea over the entire planet. It carries five times as much water as its nearest competitor, the Congo, and twenty times as much as the Mississippi River. Except for the spectacularly steep eastern slopes of the Andes, most of the Amazon Basin is fairly flat, so that much of the land is a permanent or seasonal marshland. Outside of the frozen polar regions, half to two-thirds of the fresh water on the Earth is present in the Amazon. This vast amount of water is increasingly polluted with arsenic, mercury and other highly toxic substances from mining and smelting. In the 1960s, untreated Amazon River water easily met US standards for drinking water; mining, industry and sewage

from millions of the region's new inhabitants have changed that for the foreseeable future.[60]

In addition, increasing pressures on the forests have led the state down the path of more privatization—which is presented, in line with the logic of Hardin's "Tragedy of the Commons" thesis (which I'll return to later) as the best answer to the question of how to *save* the forests. Antonio Carlos Hummel, head of Brazil's National Forestry Service, said at the 2010 Reuters Global Climate and Alternative Summit that the "future of the Amazon" can only come at the cost of giving concessions to private timber companies to manage the region.[61] It is estimated that within four to five years, a chunk of the Amazon the size of Virginia will have been handed over to private corporations and entrepreneurs.

Along with these developments, the United States continues to strengthen its military, economic, and scientific connections with Brazil, underscoring all these actions by defining them as the "improvements" brought about through the act of "freeing trade." This idea of "improvement," as we have seen, arises out of the discourse of enclosure; it was a crucial term in the discourse's new lexicon, a terminology deployed with increasing repetition from the seventeenth to the nineteenth centuries in England and then Europe. It was later applied in discussions concerning the raw material of England's and Europe's colonies from the nineteenth to twentieth centuries. It continues to be a central organizational trope used repeatedly in today's neoliberal arguments for development. The Cardoso (1994–2002), the Lulu (2003–10), and the Rousseff (2011–present) governments all embraced forms of neoliberal structural improvement. Because of these actions, the MST has had to face reductions in public services and goods and the difficulties of maintaining educational extension projects with universities and agronomists throughout each of these regimes.[62] "Freeing trade" with Brazil is equally important to the Obama administration, and to accomplish this act of freeing, the administration works with security society think tanks such as the Heritage Foundation, which has a history of pushing for closer economic-military ties with countries such as Brazil—America's tenth largest trading partner.[63] This encouragement is also an attempt to steer Brazil away from China, currently its largest trading partner. If the Obama administration is successful in this maneuver, it would raise the political, military, and economic stance of the United States in its attempts to maintain supremacy in a multipolar world. These connections are enmeshed with scientific developments in high-tech machinery, aircraft, bioresearch, and other forms of technology. At stake are billions of U.S. dollars and the environmentality of the demand for energy security in the form of hydropower, ethanol gasoline mixes, the Amazonian reserves (minerals in addition to trees), and the potential to tap into offshore petroleum reserves of the Santos Basin region.[64]

Like colonial authorities of the past, energy security advocates train their lenses on the raw materials available in developing countries. One of the most significant forms of raw material being enclosed today is sugarcane, because sugar is one of the primary substances used in alcohol-based fuel systems (the biofuels of ethanol and methanol). Lands with a good potential for the development of sugarcane crops, such as Brazil, become ideal sites for enclosure on the pretense of finding ecological alternatives to fuel. Energy security thus touches directly on environmental and human rights issues. For neoconservatives such as Zubrin interested in increasing energy security, Brazil is a "global energy giant."[65] Moreover, despite the exponential efforts of environmentalists and the military (and even a growing number of neocons) to challenge the supremacy of oil, the country is also being targeted as one of the future top-five global oil producers (this is expected to be a reality in 2020).[66] Brazil's ex-president Luiz Inácio Lula da Silva referred to the discovery of oil in the country as "a second independence for Brazil."[67] British multinational oil and gas company BG discovered the reserve in 2006, and in 2008 Brazil's Petrobras discovered a second reserve, the Jupiter field, situated 23 miles east of Tupi and roughly 180 miles off the coast of Rio de Janeiro.[68]

In April 2010, Brazil and the United States signed the Defense Cooperation Agreement (DCA). Essentially a hemispheric security pact, the agreement enacts the following: cooperation in research and development, logistics support, technology security, the acquisition of defense products and services, technological information exchanges, military training, the exchanges of instructors and students from defense institutions, and more.[69] Brazil has since signed similar contracts with the United Kingdom and Israel. These three actions helped solidify the connective tissue of the globalized security society, adding already to Brazil's 2005 and 2008 contracts with Russia and France, respectively. For these reasons, it is not surprising that the United States is now in the process of ramping up its military ties with the country. U.S. Army General Martin Dempsey met with Brazilian military leaders on March 29, 2012, a few days before President Rousseff met with Obama in the White House.[70] The goal of these meetings was to extend the economico-security interests of the Obama administration, which Obama first discussed in his trip to Brazil in March 2011.

In these meetings, Obama—stating his "enthusiasm" for Brazil's economic rise—became an emissary of neoliberal policies. He encouraged Brazilian companies to strengthen their activity with the World Bank and emphasized Brazil's need to extinguish "obstacles in the way of doing business" if it expected to become more economically productive.[71] These connections were deepened in April 2012 when President Rousseff met with Obama at the White House. Three key issues were discussed: Brazil wanted the United States to support its bid for a seat on the UN Security Council, both wanted to establish strategies to generate new economic and commercial developments, and the United States wanted both nations to move toward an equivalent military geopolitical stance.[72]

Energy Security

Today, enclosure has expanded from being the concern of a single nation or a set of nations. From the standpoint of neoliberalism and national development, the primarily agricultural concerns of the eighteenth and nineteenth centuries have now acquired broader ecological significance as substantially more resources have been targeted. The term *enclosure* is rarely used by economists or governing authorities (though it still remains a key critical term for scholars); instead, the movement has evolved into the national and global priority of a broad spectrum of "energy" and "security" issues. As governments gradually reorchestrate their modes of governmentality toward ecological issues of crises (climate change being the most pressing), these tactics begin to reveal their intimate associations with military formations of production. In the age of environmentality, the fate of the environment—biodiversity, endangered habitations, species extinction, indigenous forms of sustainability (and, in some cases, unsustainability)—and the struggles of the world's poorest communities working in cash-crop regions are being overwritten and erased by the single military allotrope *energy security*.

As perhaps the most important global phenomenon of the twenty-first century in relation to climate change, energy security signals the ultimate fulfillment of the enclosure movement. The discourse of energy security dates from the early 1970s, though its firm connection to environmental matters, specifically climate change, and its direct relation to the military have only become transparent in the years following 9/11.[73] The years immediately following the War on Terror and the publication of the Intergovernmental Panel on Climate Change's (IPCC's) Fourth Assessment Report have seen a marked increase in the use of the term *energy security* and its sister terms *energy independence, water security,* and *food security*—especially by neoliberal organizations such as the World Bank, the International Monetary Fund, and the World Trade Organization. The *Journal of Energy Security* was launched in 2008, followed by *Food Security* (2009), *Agriculture and Food Security,* and *Food and Energy Security* (both 2012). In 2007 Congress passed the Energy Independence and Security Act of 2007. The bill was originally designed to promote alternative forms of energy. In its final form, it endorsed primarily automobile fuel economy and the development of biofuels—two concerns that center on the "delinking" of the United States from its dependency on foreign ("Islamofascist") sources of oil. In 2008 a group of ecologists and environmentally minded anthropologists and political theorists working directly with national security organizations began to recast ecosystems, their entire histories, and their complex interrelations, as if they were fundamentally related to and only accessible epistemologically within the twenty-first-century security society paradigm.[74] These attempts, in other words, fully manifest what had been lurking within the discourse of enclosure from its beginnings in the precolonial and colonial eras: the desire to make security the ontological guarantee, the fundamental truth, of Nature itself.

The transformation of the environment into a "natural resource" needing to be "freed" and "secured" in the modern era was operative when enclosures became an essential political discourse of the nation-state (when enclosures, as we saw earlier, were adopted as essential to the simultaneous increase of national production and of national agricultural/land security). Energy and security had been minor concerns since the Industrial Revolution, but it was not until the Arab oil embargo of 1973 (after a thirty-year period of relative abundance and cheap prices after World War II) that nature as "energy" came to be intentionally connected to the idea of national security in the United States. It was this event, it should be remembered, that generated the politico-economic coalition that came to be known as the G5 and later the G8 governments. The G5 and the G8 recognized that they were nations that consumed the most and needed heavy amounts of energy, but they are also nations that did not have adequate domestic supplies of energy to meet their needs. The colonial legacy of enclosing the raw materials of distant territories (that only consume a relatively small amount of these materials) for purposes of heavy consumption in domestic markets has come to the point of haunting the Western and Northern supremacy of the neocolonial era. Energy products are now the "world's largest traded commodity."[75] The oil crisis of 1979, the Gulf War in 1991, and then 9/11 led to the full development of a clear discourse of energy security driving foreign policy.[76] The link among the environment (energy resources), nationality, and conflict is thus tied together in a tight constellation with the fantasy of security. In this sense, today's commitments to an environmental "cause" are becoming increasingly patriotic: finding alternatives to petroleum-based fuels, for instance, unchains us from our "dependence on foreign oil" and, by extension, foreign governments "that support terrorist organizations."

Energy security, moreover, has developed into a widespread neocon discourse in the United States since the attacks on 9/11, and it is now firmly embedded in neoliberal politico-economic agendas. Its custodians include R. James Woolsey (former director of the CIA), Frank J. Gaffney (founder and current president of the Center for Security Policy), Anne Korin (codirector of the Institute for the Analysis of Global Security and cochair of the Set America Free Coalition), Caroline Blick (senior Middle East fellow at the Center for Security Policy in Washington), Timothy Connors (director of Manhattan Institute's Center for Policing Terrorism), Cliff Kincaid (president of America's Survival Inc.), and Robert Zubrin (former senior NASA engineer), to name just a few. It is supported by global organizations such as the World Bank, the World Trade Organization, and think tanks like the Woodrow Wilson Center.

In opposition to efforts that address the causes of ecological decline and ruin and to efforts that pass meaningful legislation designed to eradicate these causes, the growing militarized trend steers ecosystems toward high-yield production, now directly understood to be a necessity because of consumer demand. It is not an environmentalism but rather an environmentality that leads to the very forms of insecurity it ostensibly avoids. To place this in the terms developed by Agamben in his analysis of post-9/11 increases in security (what he calls "the state of exception"), the privileged global class of consumers stands at the productive end of a generalized "biopower" network; they are endowed with life (what Agamben calls *bios*) and charged with the production of life. As a result, the environment becomes "bare life" (*zoē*). Maintaining the "biopower" of the privileged consumers (those endowed with *bios*) in the United States in the twenty-first century means that the state must more actively secure the (global) territory of its energy reserves, which also means that these strategies can no longer be disconnected from environmental considerations—that is, from an environment transformed into "bare life" for the sake of anthropological ascendency. As I mentioned in the introduction, this form of environmentality—the transformation of ecosystems into "bare life" resources—informs the reenvisioning of the environment as a potential target of terrorism,[77] and it enables the conservative camp to form coalitions with bleeding-heart liberals: "being green" is suddenly good politics for both sides, since both can take a hardline stand against oil producers in the Islamic world that support terrorist groups and their ideological agendas. Being green is anyone and everyone who wants to help free America and the civilized world from dependence on foreign oil.[78] As R. James Woolsey phrases it, environmentalists today are "a coalition of tree huggers, do-gooders, sodbusters, cheap hawks, and evangelicals."[79]

These maneuvers make the security state's militarizing of the environment a subject for postcolonial concern. Encouraging the use of ethanol as an alternative source of fuel, Green Patriot organizations have begun to reorient their focus to Latin America rather than the Middle East. To quote the president of the Center for Security Policy, Frank J. Gaffney, "In Brazil, at least 25 percent of the fuel sold in gas stations is sugar-based ethanol. . . . Latin American and Caribbean countries like Guatemala, Panama, Trinidad and Tobago, Costa Rica, El Salvador, and Jamaica are all low-cost sugar cane producers. These nations could become key to US energy security if large numbers of American vehicles [made the transition to ethanol]."[80] The subtext of this statement should be made as transparent as possible, especially for environmentalists looking to find ways to develop alternatives to oil. The environment is made visible in terms of its ability to yield energy, and this energy must be guarded as a matter of national security. Through its reconfiguration in this new order, environmental activism becomes a government-administered environmentality that has very little to do with any kind of familiar ecological ethic. This neoliberal mechanization of the earth and its resources further changes the nature of the colonial domination of planetary ecosystems by reinserting them into the new geopolitics of diminishing resources. "Energy security" rewrites the environment as a *paramilitary matter* and absorbs its potential as a resource not only for the biopolitics of national power but as *the* key element in international affairs. As the growing number of science, military, and national/international policy conferences taking place in Washington confirm, the neoliberal transformation of the ecosystem into energy requires the creation of a new collective international armature to ensure the continuing monopoly of resources and environments. As Jan H. Kalicki (counselor for the Department of Commerce and for Chevron) and David L. Goldwyn (former U.S. assistant secretary of energy for international affairs) argue, "Collective energy security in the twenty-first century, like collective military security in the twentieth century, can transform competitors into partners and allies." The U.S. Department of State has recently begun to mirror this collective energy security policy, creating a series of six international partnerships, key among them being the China–U.S. joint EcoPartnership.[81]

This new move to secure the environment is a multistage process, overwritten by neoliberal policies of "opening markets" to the rewards of development. First, advocates of energy security argue for the "emancipation" of all land resources in developing countries. Emancipation is a polite neoliberal name for the privatization/enclosure of common or indigenous land and its resources by a powerful, global conglomerate. Indigenous and local communities are charged with "protectionism" if they attempt to resist such acts of "resources extraction." The World Trade Organization perceives agricultural protectionism—an essential aspect of communal land organizations such as the MST, which I will turn to in the last section of this chapter—as the biggest obstruction to development. The World Trade Organization's Doha Ministerial Declaration, adopted on November 14, 2001, lays the groundwork for the destruction of alternative land movements struggling to maintain a political presence in agricultural communities around the world. As Joseph Stiglitz and Andrew Charlton have argued, "We've settled on a program that lays out ambitious objectives for future negotiations on the liberalization of the agricultural market. These objectives . . . will help the United States and others to advance a fundamental agricultural reform agenda."[82] World Bank lead economists Kym Anderson and Will Martin claim that developing countries will "enjoy 45 percent of the global gain from completely freeing all merchandise trade."[83] In addition, developing countries are encouraged to expand their infrastructure to prepare for the oncoming demand of high yields from developed countries.[84] In recent years, Brazil has been of special interest to the United States in this respect.

This constitution of panoptic land is now in the full stages of planet-wide development. This is being reflected mostly by the United States and international development agencies, where, despite differences in historical development, the discursive imperatives of improvement, increased yields, efficiency, entrepreneurship, and security continue to govern the enclosure and privatization of environments.[85] Kalicki and Goldwyn, for instance, view the relation between foreign capital and the struggle to protect the environment as a roadblock that "stalls" economic development "in much of Latin America."[86] But economic development, they argue, cannot occur without first being firmly coupled with a panoptic armature of security that touches on geopolitical regions in need of policing by the United States and its allies: "The US needs to build coalitions among nations in both the industrial and developing worlds to support a wider shared security agenda, including such issues as systemic reform in the Middle East, transparency in Africa, and the rule of law in Russia and Central Asia."[87] Its expansion has already begun to produce the "accidents" of increased greenhouse gases and increased pressure on ecosystems to generate more "energy." This pressure on ecosystems also puts pressure on the indigenous populations that inhabit these territories.

The inhabitants of key ecological regions, such as the MST and the indigenous of the Brazilian Amazon, confront two forms of opposition: the elite of their own governments and the foreign governments interested in "opening up" their habitations to gain access to its "energy reserves." With growing frequency, these national and international forces are backed by a strong and unforgiving military armature. National elites talk to and establish developmental connections (in the form of administrative legal documentation and in material technological developments in specific regions—dams, mining complexes, agribusiness, logging) with global forces of neoliberalism (the World Bank, transnational corporations). In the case of the Amazon, for instance, the Brazilian state and military work directly to advance economic interests; indigenous territories are subsumed into programs of energy exploitation. When concerned environmentalists from within the nation and environmental organizations of the global community outside the nation of Brazil criticize these as ecologically destructive practices, the Brazilian state reframes and deflects this criticism as an example of "international conspiracy" that threatens "national security" and the inherent right of a nation to maintain its sovereignty.[88] This idea—that outside forces are only ever interested in threatening the sovereignty of a nation and its inhabitants—filters down through the population, even at times affecting those indigenous inhabitants that seek to oppose neoliberal developmental models.

This explains why some indigenous peoples are suspicious of foreign environmentalists and their efforts to join in their cause or speak on their behalf. Indigenous concerns are often spoken for by others, by operatives on other sides of the struggle, and complex arguments for the rights of citizenship are quickly and easily subsumed by higher-order justifications for economic interests (this "higher-order" mechanism is a phenomenon explored in more detail in chapter 4). In addition, indigenous inhabitants are viewed by both national and global neoliberal forces as enemies of economic growth and an unwanted barrier to development in all its forms (infrastructural, technological, social, cultural, intellectual, etc.). This view is further complicated by military forces and nationalists that see *any* discussion of environmental preservation to be a threat to national security. At the time of the 1992 United Nations Conference on Environment and Development (UNCED-92), held in Rio de Janeiro, Brazilian military leaders and nationalists suggested that the conference be supervised by "people capable of defending national interests: the Armed Forces."[89] At the same time, the military characterized outside interests as a threat to not only the Amazon but indigenous inhabitants as well. In this sense, the military, nationalists, and transnational corporate actors seek to monopolize the development of the Amazon and dismiss international cries about the environment as a massive conspiracy to mystify what they argue to be the real agenda behind environmental "concern": that outside forces only want to control for their own selfish gain the development and "improvement" of the Amazon.

Indeed, given Northern interests in the Global South, such criticisms of international efforts to enclose are not unwarranted. This is epitomized by Robert Zubrin in his book *Energy Victory: Winning the War on Terror by Breaking Free of Oil*. In his book, Zubrin essentially makes the case for a total, panoptic enclosure of practically all third-world agricultural communities. He espouses the logic of "improvement" that pushes for higher yields that exemplified the writings of proenclosure advocates of the eighteenth and nineteenth centuries:

> The United States imports 12 million barrels of oil per day, more than a third of the OPEC total. In 2006 this oil cost us about $260 billion. If we were to spend this on alcohol fuels instead, we could increase the *total* US farm income of $127 billion by 50 percent—and still have around $200 billion left over to pay for alcohol fuel imports derived from third world agriculture. For comparison, US agricultural imports in 2005 totaled $56 billion. Thus, by switching to alcohol we could *quadruple* our purchases of third world agricultural goods, while giving US farmers substantially *more* business, not less. . . . If the pattern described above were followed by other oil importers as well, the net effect would be to multiply total third world agricultural foreign exchange earnings fourfold, yielding earnings that would dwarf all current worldwide foreign aid expenditures by a factor of ten. . . . A huge engine for world development would thus be created.[90]

Zubrin's book makes England's development of panoptic land look like child's play. His argument for developing the potential of methanol is based on the fact that the crop base for producing methanol includes "all plants, without exception": "Not only the edible and inedible parts of commercial crops, but weeds, wild jungle underbush, trees, grasses, fallen leaves and branches, water lilies, swamp and river plants, seaweed, and algae, can all be used to produce methanol" (152). He essentially views the world's many "developing" countries—especially Latin America—as essentially America's farm: "It would behoove us to create jobs south of the border by buying much of our methanol and ethanol abroad" (155). His argument for the foreign development of ethanol and methanol precisely replicates England's relation to its former colonies by maintaining colonized spaces as sites of raw materials and by ensuring that the host country controls the means of production: "Put money in their hands that they can use to buy our manufactured goods, such as tractors and harvesters to grow the fuel crops, trucks to transport them, and equipment to process the biomass into liquid fuel" (156). Indeed, Zubrin even devotes a chapter of *Energy Victory* to the rise of the alcohol fuel industry in Brazil;

he presents Brazil's military dictatorship as its savior: "When the newly inaugurated General Ernesto Geisel found his nation confronted with economic devastation from the Arab oil shock, he acted forcefully to seize his moment for greatness. Even though Brazil imported 80 percent of its oil, the enlightened military autocrat decreed that it could and would become energy independent. His answer was ethanol" (162). Geisal's institution of low-interest loan programs fronted the construction of hundreds of new ethanol distilleries. He then mandated that the alcohol ratio in flex cars be at least 24 percent. Consequently, "ethanol production soared 500 percent in four years" (163). (Geisal's program was not the miracle that Zubrin paints it to be. During this period, Brazil remained financially insecure because of accelerating international interest rates and a slow growth in world exports.)[91]

The important point here is that Zubrin's plan seeks to improve the efficiency and high yield of all agricultural regions while at the same time firmly connecting that improvement to the United States and its corporate allies. Despite the direct relationship between ethanol production and the growth of devastating poverty, Zubrin presents the Brazilian experience as a triumph of Man over Nature that "should be copied" by "many in the developing world."[92] In fact, Zubrin sees this transformation as a military necessity. He argues that the entire region of Latin America should be transformed into an ethanol farmscape so that it can ward off the spread of "Islamist ideology": "While America sleeps, Islamists are using their alliance with Latin American drug lords to establish bases in the Western Hemisphere" (175).

Crucial to an understanding of Zubrin's espousal of a panoptic enclosure of land for purposes of inaugurating a great new age of biofuels is his reference to the "correct" way to govern economically and politically. It is not by accident that his referential point is England: "In European history, the acquisition of vast unearned New World treasure by Spain led to the degradation of the enterprising classes and caused the abortion of the development of representative institutions in that country. In contrast, the need of England to develop its sources of revenue through commerce and manufacturing forced the British monarchy to consult with Parliament and enact measures suitable for growing the productive powers of the nation" (251). But neither the environment nor the land nor the people who work the land are ultimately important to Zubrin and his energy-security compatriots. His supposed concern for the environment becomes clear when he characterizes Al Gore's assertion that "we must make the rescue of the environment our central organizing principle" as a stance that comes "dangerously close to proclaiming [its] loyalty" to "fascism," "totalitarian communism," and "Islamism" (259). Zubrin's simultaneous espousal of parliamentary "productivity" and condemnation of Gore's politicizing of the environment are symptomatic of environmentality's "postenvironmentalism" (the program espoused by Friedman, Nordhaus, and Shellenberger). A policy of enclosure thus becomes the perfect structural fantasy for obtaining an attainable level of planetary power as the waning of oil supplies looms on the horizon.[93] In both cases, habitations and their inhabitants are constituted as "bare life," having value only in their ability to supply energy for the industry of life.

As with all attempts to enclose land, the impulse for high productivity ultimately imagines the land as an enemy needing to be disciplined, which in turn leads to the destruction of that land and the lives of inhabitants attempting to carve out a meager existence in the face of overwhelming global mechanizations. What no advocate of the "Brazilian miracle" discusses, for example, is the astonishing abuse of the people who do the bulk of the labor necessary to produce the yields demanded by the international market. Brazil has one of the world's most highly concentrated landholding structures, "with a Gini-coefficient near 0.9. . . . nearly total concentration of ownership in few hands." This situation has resulted from Brazil's colonial history, a latifundio-style agriculture and land tenure system (the concentration of landholdings in the hands of a few wealthy individuals).[94] More than 50 percent of land is controlled by 4 percent of landowners.[95] Nearly a third of Brazilians, approximately fifty-five million people, live on less than two dollars a day, while another third live in absolute poverty: "By most measures, Brazil has the largest gap between rich and poor of any sizable nation on Earth" (xvi). "Experts frequently classify the poor of Brazil's Northeast as the largest group of severely malnourished and disease-ridden people in the Western Hemisphere" (xix). More than three hundred thousand farm workers are seasonal cane cutters in this "Saudi Arabia of biofuels": "By most accounts, their work and living conditions range from basic to deplorable to outright servitude."[96] In a 2008 report, Amnesty International detailed the grim situation of Brazil's biofuel workers, more than "1,000 of whom were rescued in June 2007 after allegedly being held in slave-like conditions at a plantation owned by a major ethanol producer, Pagrisa, in the Amazonian state of Para."[97]

In stark contrast, Brazilian corporate executives rank among the highest paid in the world.[98] As we saw earlier, this is due primarily to Brazil's history as a colony of Portugal and to its production of sugarcane. Sugarcane has been the dominant crop in Brazil since 1537—a time in colonial history when sugar was worth almost as much per pound as gold. Between 1580 and 1680, Brazil was "the world's largest producer and exporter of sugar."[99] The period was known as Brazil's golden "century of sugar." The production of sugarcane began in the northeast of Brazil, the area that contains more than half of the poor people in the country and two-thirds of the rural poor: "The production of sugar cane in northeastern Brazil marked the creation of a very powerful class that controlled large landholdings and exercised almost total power over both the land and the people who worked it" (107–8). Yet global organizations such as the World Bank, neglecting these local and global histories, are now utilizing the full potential of the discourse of enclosure. In the case of Brazil, the government's agrarian policy more often than not follows the World Bank's recipe for economic development.

Retrieving Inhabitancy: Gateward's Case

Up until this point, I have been using *inhabitancy* as mainly a theoretical concept, ungrounded in any historically specific occasion. However, the conceptual nature of the term stems from a specific historical event. Roughly sixty years before Parliament ratified enclosures as a national agenda, when the battle over open land was at some of its most brutal moments in an era of great impoverishment, a judicial litigation named Gateward's Case was heard at the Court of Common Pleas. Second only to the Court of King's Bench (which in some cases was less influential), the Court of Common Pleas oversaw disputes between citizens (disputes that did not directly concern the king). It primarily handled cases concerning debt and property, and it was the only court where claims involving real property could be brought. This gave the court a higher authority to set precedent over other courts. In addition, the court was where most students of law went to learn, and it shaped medieval common law more than any other British legal institution.[100] It existed for more than six hundred years until it was merged into the High Court of Justice in 1873. Sir Edward Coke, one of England's most influential judges and perhaps the most powerful judge of the seventeenth century (when the enclosure movement was transformed from an unorganized undertaking by individual lords into a national imperative), considered the court to be the "lock and key of common law."[101]

Gateward's Case (1608) centered on the rights of common land and the legal validity of the term *inhabitant*—a standard and widely used term of custom law that had a long history of being invoked to justify use rights and land tenure on common land. The key issue of the dispute revolved around a land tenure custom that had been in use for centuries known as *profit à prendre*: the use rights to soil, crops, timber, and so on and the right to access pasture for purposes of grazing cattle and sheep. These forms of open access were the right of all inhabitants living on common land. It is important to grasp the original importance of this custom, which can be misunderstood in terms of the postcustom law of today's common law. In the twenty-first century, *profit à prendre* names the right to enter *private* land and remove a product from that land. It is opposed to *easement*, which names a "nonpossessory interest" in another's land (the ability to cross another's land in order to gain access to a public road, for instance). Under the custom law/commons system, *profit à prendre* names something very different: the basic right to land for all. This basic right is the originative starting point for the constitution of the preenclosure *bios*. Under the discourse of enclosure, this right to land becomes, to new entrepreneurial ears, an irritating claim from the jungle of unwritten, vague, and bartered laws of *custom*. In the supplanting of custom law by the new common law—a document-obsessed construct grounded in privatization, individual entrepreneurship, and the anticommunal pursuits of capitalism—one's basic right to land is transformed into an act of theft.

Gateward's Case specifically revolved around the freedom to pasture cattle in an area of land that, "since time immemorial," had been considered "reasonable for inhabitants to dwell" (*ratione commorantiae*). The concept of dwelling in the phrase *ratione commorantiae* arose out of the fundamental existential *custom* act of accessing one's right to common of pasture. This right to dwell indicates a certain type of subjectivity that eventually becomes completely alien to common law and the discourse of enclosure. At first, the intensity with which enclosure advocates and common law judges wage war against the concept of inhabitancy indicates just how precisely the concept is understood—for the inhabitant is, in fact, the major enemy of enclosure and common law. The value of investigating this subjectivity, therefore, should not be found in espousing it uncritically as if it were a truer form of human existence. Its value, rather, lies in the way it reveals the violence and restrictive limits of the enclosure paradigm, which has historically been defined as the beginning of freedom for Western man (the freedom from serfdom and being "chained to the land").

The element of dwelling (*commorantiae*) as the constitutive essence of the inhabitant is obviously related to what is commonly understood in poststructuralism as the project of "contextualizing" or "historicizing" the subject, informed by the Heideggerian decentering conceptions of "dwelling" (*wuon, wunian*) and "being-in-the-world" (*Dasien*). However, the dwelling associated with the right of access to land is not exactly the same; it needs to be understood more precisely as a specific kind of spatiotemporal *act*, one that contains an element of what might be identified as "holding space" or "accessing space." *Commorantiae*—"remaining," "lingering," "detaining"—is the plural participle of the verb *commoror*: "I stop," "I stay," "I remain," "I linger," "I delay," "I detain," or "I am detained." Being constituted as a being-in-an-act-of-dwelling, since the key here is the performative act of dwelling, and identifying dwelling in terms of "staying" and "lingering" embody the condition for the possibility of the inhabitant. But this "holding" of space is not the enclosing of space. Each of the terms—"delaying," "lingering," and so on—tacitly indicate a certain temporality that marks itself as nonpermanent: "linger [for a while]."

This lingering refers, of course, to the practical needs to acquire one's sustenance. But this is not a simple "of course" and should not be seen in reductively pragmatic terms. Dwelling as lingering names the nonaccumulative (nonprofit-driven) sustaining of life in an open-field agricultural community. It is not "bare life" (*zoē*) in the sense of a life denied the basic rights to existence and capable of being eliminated without legal or political redress, nor is it bare life in the sense of a life abused in order to serve the life of another. *It is the basic, qualified political form of life* (bios) *defining human agricultural-environmental existence in the era of the commons.* In the wake of enclosure, "lingering" will be negatively recast as "loitering." The basic form of custom-law life will become essentially an illegal act, one understood to be at war with the rights of the enclosing landlord. It will come to signify the unlawful act of trespassing (on an individual or corporation's property) and indicate as well an inappropriate use of time (in the sense of "dawdling" or "loafing"). But in the system of custom law, "lingering" signifies something entirely different. It refers to the act of maintaining one's life. In the dwelling-staying-lingering action associated with inhabitancy, we find the *bios*—the qualified given life of the inhabitant and the sustainability of having access to a livable habitation. It comprises an activity not bound to an overarching system of rules and regulations. It is instead connected to local heterogeneous relations that constitute a state of existence that has yet to be subject to a generalized economy of science (economic or political) or a normalized state of governmentality.

During the case, Gateward pleaded according to custom law: that dwelling in the common had been the practice of inhabitants since time immemorial. However, it was unanimously agreed by all the justices of the Court of Common Pleas that the custom—and by extension the entire history of custom law—was unlawful. The reasons are detailed and listed in the court document:

1. Only enclosed spaces, or commons associated legally with enclosed spaces, are recognizable as legal spaces.
2. Only a man with an interest in a house associated with a former common (now enclosed) can have interest in a common.
3. "A custom can only extend to that which hath certainty and continuance."
4. Inhabitancy "would be against the nature and quality of a common, for every common may be suspended or extinguished, but such a common [a space open to all and owned by no one] would be so incident to the person that no person certain could extinguish it" (in other words, if they allowed Gateward to win on the grounds of custom law, which was now resignified as vague, it would be impossible for any landlord—who was given "certainty" by the new common law—to extinguish/enclose a common).
5. An inhabitant can have no claim to a common.
6. Because of 5, an inhabitant has no "sufficient interest" that would enable him to plead in a legal proceeding.
7. "No improvement could be made in any wastes if such a common should be allowed"[102] (in other words, a common—by definition subject to the constant traffic of all—would never be free for improvement).

The convoluted language of the rationale, especially in reasons 2, 4, and 7, is indicative of the uneasy relationship the new, developing common law must maintain as it enmeshes its early forms in the spaces of everyday social existence still operative in the waning political life of custom law. The goal at this early developmental stage is to eradicate not only the justification custom law gives to the open spaces but also eventually open spaces entirely. Yet it must necessarily face the continuing presence of open habitations and their inhabitants. Hence, for instance, the repetition of "such a common"—an indication of the in-between status of common spaces: the common is there, yet it should somehow be different from what it was before, not entirely open and regulated according to the jurisdiction of those associated legally with enclosed spaces.

These highly detailed restrictions constitute a hinge moment in the war against inhabitancy, custom law, and the open access of the commons system. They symptomatically reveal the powerful legacy of the commons system in and through the convoluted nature of the restrictions. The common is clearly still an existing open space (it is not enclosed), yet it is also no longer to be accessed as a fully open space. The court cannot deny that it is not enclosed, but it can also not let it be fully open anymore. More to the point, the common is only to be open to an *entrepreneur* who owns an enclosed plot of land. In this precise manner, both the *inhabitant* and *habitation* are being targeted. Erasing dwelling/habitation from existence erases the subjectivity of the inhabitant from existence as well. Each point of contact between the inhabitant and his or her habitation is being severed.

Gateward's Case and its subsequent citation and influence in future cases constitute the key legal mechanism in what will eventually become the state's adoption of the enclosure movement. If common spaces are the only spaces made available to inhabitants, then erasing access to common lands constitutes the complete erasure of the subjectivity of the inhabitant. Only one form of access to a formerly open common was allowed: inhabitants could have access to the land in question if they needed to cross it on their way "to the church or market, etc."[103] But this "easement" is not an act of habitation; it names the passage from one legal (owned) space to another, like the crossing of an ocean in order to leave one continent for another.

According to nineteenth-century barrister and legal historian Thomas Arnold Herbert, Gateward's Case became the "leading authority for the proposition that inhabitants unless incorporated [that is, unless they can be recognized in common law as having a certain connection to the land], cannot by custom or prescription establish a right to any profit *in alieno solo* [on the property of another]."[104] The number of cases that referred to Gateward's Case as precedent are considerable, and in each of these cases, the thrust of the decision was centered on the term *inhabitant*, which was judged to be, at best, a weak claim to the land: "*occupiers of lands may by custom claim a right in alieno solo; though inhabitants cannot*, because inhabiting is too vague a description, and extends to many others besides the actual occupiers of houses and lands."[105]

Prior to Gateward's Case, invoking the term *inhabitant* was enough to overthrow a lord's act of land grabbing from a community of farmers. It spoke of the imminent relation between the human and an environment. Existing in a "habitation" was not yet a right that one had to demand or protect before a court of law; the concept was part of the deep understanding of what it meant to be human. Gateward's Case essentially changed the ontological constitution of human subjectivity, declaring that beings constituted in relation to a habitation no longer had a right to exist. Such beings were shifted to an undefined status: "The court declared that rights of common could not be claimed in the name of a group as amorphous and 'uncertain' as the 'inhabitants' of a district."[106] Over the course of the next two centuries, the judicial system would devote its efforts to the erasure of custom law by going to war against inhabitants, ensuring that the concept of the human as an inhabitant was struck from the judicial register and, by extension, sociopolitical existence.[107]

Far from being an isolated event, Gateward's Case constitutes a sea change in the development of the modern legal system. By the very fact that this event and the idea of inhabitancy have been practically buried in the historical record is, perhaps, an indication of its success. It serves as the lynchpin of the soon-to-explode world of imperial colonization of foreign land, which, if anything, is about the gradual erasure of indigenous rights to inhabit unenclosed land. E. P. Thompson, in his analysis of the invention of modern-day "common law" and its war against ancient laws of custom, argued that the case's erasure of the rights of common and its imposition on traditional open agrarian communities was instrumental in the establishment of new capitalist definitions of property.[108] Custom law was found to be "unreasonable" to the new legal instrument of class expropriation orchestrated by judges, who had vested interests in maintaining their status as landowners. Thompson argued, in fact, that this single case had more of an effect on the destruction of custom laws than any other event. The ruling of Gateward's Case was invoked throughout the eighteenth century and used as a rationale for the gradual "destruction of custom law and the loss of long-standing use-rights of the landless cottager."[109]

Gateward's Case was a decision made out of common law's desire for calculative control, for it made possible the imposition of greater constraints on the activities of the national community. According to legal historian Andrea C. Loux, the case was the embodiment of a "newly declared element of certainty" that "had a dramatic impact on customary use-rights."[110] The highest court in the land—the Court of King's Bench— was instrumental in applying Gateward's Case and, along with it, this new logical economy of certainty, which became the grounds for a new legal rationale:

> Prior to the establishment and application of the four-part test in the early seventeenth century, reasonable custom was recognized and applied by the common law courts. The development of a new judicial test limited the extent to which local custom was judicially noticed as operative law. The test also regularized customary use-rights by attaching those rights to recognized tenurial interests in land. Aside from the doctrinal significance of Gateward's Case, it is also an important indicator of the effect of social forces on the development of customary law. The court would not recognize a right of common vested in all inhabitants because a common that could be claimed by virtue of inhabitancy would be "transitory and altogether too uncertain, for it will follow the person, and for no certain time or estate but during his inhabitancy, and such manner of interest the law will not suffer, for custom ought to extend to that which hath certainty and continuance." Although the court expressed its refusal to recognize the inhabitants' right of common in terms of certainty, its chief policy concern was that landowners be able to enclose. The court noted that "every common may be suspended or extinguished," but if the right of common were vested in inhabitants, rather than copyholders of the manor, "no person certain can extinguish [the common]."[111]

This insertion of certainty into the judicial register of human sociopolitical forces of production should not only be seen as a demand of capitalism; it must also be understood against the larger backdrop of increased governmental demands to put into operation a general security of the land.

In point of fact, Gateward's Case would not have had such an extensive effect (and, in fact, might have been lost in the endless miasma of legal history) if it were not because of the efforts of the legal giant Sir Edward Coke, often called the greatest of English judges. As a leading figure of the new, fledgling biopolitical order, Coke sought to supplant the discrete and heterogeneous system of custom law with the more homogenized common law. In his efforts to confront the "lawless hoards" of England's inhabitants, he was basically making population a problem roughly a century before the historical moment that Foucault identified as the rise of the state's concern with its population. In the push to create a standardized form of national conduct, Coke was essentially advocating an early system of governmentality. He actively supported the enclosure movement and pushed for the "improvement of wastelands across England."[112] In the 1624 Parliament of James I, he and others proposed that all antienclosure legislation be repealed.[113] Though his efforts were unsuccessful at the time, they were instrumental in influencing Parliament's eventual full endorsement of enclosure during the Commonwealth period.[114]

Coke in fact had a history of dealing ruthlessly with any form of antienclosure resistance. In the Oxfordshire Rising of 1596, he "almost single-handedly controlled the prosecution of the leaders" of the movement.[115] This particular antienclosure movement was a miniscule event when set beside larger protests that occurred throughout the nation during the same and other years. Yet Coke intensified the importance of the event by turning it into an example of a major threat to national security. The rising involved a small group of starving farmers who were planning to gain the support of a few hundred men from towns in Oxfordshire to march on London. The event took place during one of the Tudor era's worst years of harvest. The rising was in protest of enclosures and the increased holdings of large landowners, which were occurring at a time when 20 percent of the population of England were living below the poverty level. The leaders of the movement failed to gain enough support, and a small group was arrested before getting to London at the hamlet of Enslow Hill. Coke aggressively pushed for the leaders of the protest to be brought up on charges of treason, against the misgivings of several other judges. He cited precedent for the charge by referencing a 1571 statue of treason, despite that the leaders' actions had not met all the requirements of the statute. The men involved in the rising were tortured under the suspicion that they were involved in a wider conspiracy against the queen and the nation. The two leaders of the rising were hanged a year later.[116]

During the hearing of Gateward's Case a decade later, Coke argued that custom had no true lawful origins, because custom was by definition a form of agreement that opposed the more universal authority of common law. As we have seen, the concept of inhabitancy was grounded in the nonuniversal heterogeneity of custom and the belief that customs were distinguished from common law by the fact of their local provenance. Tellingly, one of the few moments when custom had to justify its existence outside of its immanent context occurred in the act of making reference to its practice as having been operational "since time immemorial," whereas common law, which was a new homogenizing regulatory apparatus attempting to spread itself across the realm, never had to make recourse to such a justification. Its justification lay entirely in its own highly documented and calculatory representational apparatus, which in turn was distributed and given a regulatory dimension by an expanding administrative machinery. Common law, in other words, depended on its self-constituted discursive formation of panoptic land. To understand how custom could involve practices of inhabitancy across the land, it must be remembered that customs were not an invention of the nation-state, which came into existence as a fantasy in the wake of such nationwide circulatory devices as the newspaper. Such devices, as Benedict Anderson has powerfully argued, were instrumental in the formation of the nation as an "imagined community"—a community made possible in and through the existence of homogenizing, panoptic narrations. Custom, on the other hand, arose from local collective agreements that needed no globalized justificatory apparatus. And these agreements varied. Such variations were possible within the openness of the commons system. An antisystemic form of convention of this type could not have originated within the highly organizational demands of a system such as that of enclosure, which does not recognize the heterogeneity that arises from more immanent relations.

The greatest waves of enclosures occurred during the eighteenth and nineteenth centuries, making the normalizing power of the panoptic enclosing apparatus still somewhat developmental and incomplete. The system still had to confront customs of inhabitancy being maintained during the latter half of the nineteenth century. In an odd turn, these customs were sometimes recognized and even continued to exist, but only through the creation of a peculiar fantasy in which the origin of these customs lay in a more empirically verifiable—and thus trustworthy—foundation. According to Andrea Loux, the nineteenth century created the mythology that such customs were granted by feudal lords, and not in unknown, "immemorial" origins. This coupling of custom to the new demand of certainty and not the immanence of custom law in the form of positive legal rules enabled the custom of inhabitancy to maintain a meager form of existence before it was completely erased by common law and the discourse of enclosure. If a custom could be shown to "coincide with some recognizable jurisdiction"—a specific district and a specific manor house—then it could ostensibly be anchored to a dominion associated with an individual, not the vagaries of communal, immemorial praxis. But this legal positivism only worked in some cases against enclosed land and only applied to copyholders, not to completely landless inhabitants. For most inhabitants, "the courts, having created an historical fiction of the democratic feudal manor, disallowed custom when it could not have originated by grant of a feudal lord."[117] In the seventeenth century, it was well known that the origins of custom did not lie with the feudal lords. The fantasy that custom could be linked to contemporary positivistically verifiable origins had yet to be invented. During the rise of common law, it was crucial that custom reveal itself to be an unreliable form of social coherence. This was why, for Edward Coke, the new legal system needed to emphasize in no uncertain terms that immemorial customs had no true origin in certainty. In such a way, the nonverifiable origins of custom could be always already "lost to history" and, therefore, not properly understood to be legal forms of existence at all.

As the discourse of enclosure and its enforcement of privatization as the basis of existence expanded into full-fledged capitalism, the full range of open-access existence became increasingly reterritorialized and resignified as profits that could not be claimed by inhabitants. According to the newly invented legal rationality, the constitution of such a non-property-oriented subjectivity not only had no legal right to such claims but also could not be recognized on a fundamental level as a proper form of human existence. Customary understandings of a livable existence based on *profits à prendre* continued to be abolished throughout the industrial era, thus facilitating the expansion of a market economy. The new capitalist economy began to justify its existence by marking itself more and more as a form of "freedom" opposed to the "servitude" of the old feudal system. The stark servitude of the feudal era is a point that, of course, Marx himself emphasized throughout his many discussions of precapitalist economic formations: that "serf labour was chained" to "landed property" during the feudal era.[118] In fact, Marx was instrumental in the creation of a general historical understanding of the era immediately preceding capitalism. His insistent characterization of the feudal era as another form of exploitative social relations is important to acknowledge, for it was certainly not a golden age of human existence. Nonetheless, it is curious that Marx never discussed or even touched on inhabitancy as a significant element of precapitalist social relations, especially given the extensive legal discussion of inhabitancy and the protracted judicial war against the very term *inhabitant*. It is certainly true that feudalism contained its own form of servitude and restrictions of human freedom. Yet as we've seen, the concept of the inhabitant does not owe its existence to feudalism. Its origin, though unclear, is to be found in a different paradigm—perhaps in the set of nomadic relations that the great agricultural historian Joan Thirsk suggests might be the original impetus for rotating crops. It is difficult to unearth the precise ontology and sociopolitical nature of inhabitancy, partly because it was not of a system that called for extensive record keeping. As a generally nonwritten phenomenon, we can at best only surmise what the act of inhabitancy entailed by reading the struggles documented in court proceedings and in the generalized statements demonizing the practice by the custodians of enclosure. Because of this, inhabitancy exists in a kind of historical abyss that is not fully accessible from the standpoint of the modern, postindustrial era.

Feudalism, it should be noted, was not as extensively record and map oriented as the enclosure movement. Nonetheless, its central exploitative mechanism, the tithe system, required at the very least a functional documentation system. This came in the form of the "tithe apportionment"—which contained rough maps, descriptions of land, states of cultivation, and rent charges. Though it may not have matched the documentation standards that arose in the seventeenth and eighteenth centuries with the enclosure movement, feudal and aristocratic records give a clear picture of feudalism's management methodology. As such, "feudalism"—defined as an exploitative mechanism made up of serfs chained to the land by a dominating aristocracy—could serve as the perfect enemy for the discourse of enclosure and the political economy of capitalism. This postfeudal wholesale characterization of history before capitalism conveniently subsumed all previous conceptions and performances of inhabitancy under the general heading of the privative signifier "serfdom." Thus postfeudal legal rationality orchestrated its efforts around a fantasy that substituted the enemy of feudalism in place of its true other: "The presumption of a custom's origin in the grant of a feudal lord was the creation of eighteenth and nineteenth century judges who sought the origins of custom in a mythic feudal past. In earlier centuries, courts acknowledged that the origins of certain customs had been lost to history."[119] (It should be remembered too that there were also villages existing outside of any feudal lord's reach.) In this way, a very different alternative to the new political economy—an alternative that had flourished for at least six centuries— was radically erased from the very domain of historical consciousness.

It is important not to advocate the alternative of custom law and inhabitancy as some idealist dream. There were many existing disputes over territory and inhabitancy prior to an act of enclosure, as early enclosure award documents indicate (these documents extensively detail the status of communities' social-territorial relations in an attempt to resolve all conflicts before the enclosure surveyors parceled out the land). Rather, the point is to symptomatically interrogate the way in which the discourse of enclosure went to war against the idea of inhabitancy: its intensity in dismissing inhabitancy's viability as a legal form of existence; its dismissal of inhabitancy as lacking any architecture for a sufficient form of governing; its condemnation of the inhabitant as an idle and obtrusive form of being human; its eventual association of inhabitants with "savages," "gypsies," "vagabonds," and "masterless men"; its anger at an ecological method of existence that did not see the land as a wild entity in need of securitization, extensive improvement, and cultivation; and ultimately and most importantly, its dread of a form of life not based on capital accumulation.

Inhabitancy and the Example of the Brazilian Landless Workers' Movement

Despite inhabitancy's nonrepresentational nature, the original tensions that surrounded the war against this form of subjectivity continue to operate in the present, generating an untimeliness in the general neoliberal schema of existence that deserves attention. If we consider the impact of the legacy of the enclosure movement on present-day neoliberalism and its coconstituent militarization of the environment, we can make a strong case for identifying inhabitancy as capitalism's true "other." Inhabitancy is the specter that haunts capitalism or, to speak like Žižek, the impossible Real of capital. As the Real, inhabitancy has no recognized place within the system, which is why it appears in the form of a struggle. Though formations of inhabitancy in the postcolonial era contain obvious historical and geographical differences from the midmodern period of enclosures, twentieth- and twenty-first-century contestations for noncapitalist human–land relations nonetheless bear considerable resemblance to the phenomena of the seventeenth and eighteenth centuries. Organizations such as the Communal Property Association of South Africa, La Via Campesina, the Rural Women Workers' Movement in Brazil, the Communal Irrigation Management System of the Swiss Alps, the Middle Atlas Central Area Agricultural Development Project in Morocco, the Uganda Land Alliance (and similar alliances in Kenya and Rwanda), the Dalit Movement in India, and, of course, the MST, to name only a few, struggle over issues of inhabitancy and habitation rights and confront the neoliberal world that gives greater priority to socioeconomic mobility and technological innovation than to the needs and values of communities that are used and abused by rapid developmental changes. Regardless of the differences between the seventeenth- and twenty-first-century constitutions of inhabitancy, the pressures that privatization and development inflict on habitations have increased exponentially since the days of Gateward's Case, making the issue of habitation one of the most significant problems of the twenty-first century.

It would be naïve to attempt to unearth a phenomenological presence at the heart of the concept of inhabitancy. Indeed inhabitancy appears to manifest itself as an activity or performance, as the court cases surrounding Gateward's Case and the many disputes between enclosures and commons reveal. Different geographical, ecological, cultural, historical, and political contexts are important to bear in mind. Nonetheless, we might venture forth some provisional characteristics in our exploration of inhabitancy as an experimental *concept*. Inhabitancy as a performative force appears to arise from the standpoint of a co-relation to a people and a place. As I've suggested elsewhere,[120] etymologically the term connotes a form of subjectivity borne along by an environment, an indissoluble relationship between the subject/inhabitant and the land/habitation: *habiting*, from the French *habiter* ("to have dealings with," "cohabit," "dwell," "inhabit") and from the Latin *Habere* ("to have," "to be constituted," "to be"). The last definition from the Latin reveals the extent to which the concept connects being and existing to the act of living in a place. *Stroud's Judicial Dictionary of Words and Phrases* states that the word "has no definite legal meaning, its signification varying according to the subject matter."[121] The concept often refers to the poor, in the general sense of a set of underprivileged residents, and can be found in many references to the "poor of a community," as in the following: "A legacy for the benefit of the 'Inhabitants' of a place would seem a good charitable bequest . . . as one to the 'Poor Inhabitants' certainly is, under which the persons to be benefited are those poor inhabitants who are not in receipt of parochial relief."[122] The American judge and legal historian Benjamin W. Pope cites several definitions of the term (inherited from its judicial history in the English system) as used in court cases. He endows it with a special form of universality, one that emphasizes an essential, nonpositivist right to freedom:

> The term is conceived to be entirely free from technicality, and has a known and universally accepted meaning: all agreeing in considering inhabitant as directly connected with habitation and abode. It is supposed, that a term of no technicality, so simple and expressive in itself, and so clear and definite in its character is susceptible of but one meaning. This residence, however, is to be bona fide, and not casual or temporary. The word inhabitant comprehends a single fact, locally of existence; that of a citizen, a combination of civil privileges, some of which may be enjoyed in any state of the Union. The word citizen may properly be construed a member of a political society; and may properly be absent for years, and cease to be an inhabitant of his territory, the right of citizenship may not thereby be forfeited, but may be resumed whenever he may choose to return.[123]

Significant here is the characterization of the term as "entirely free from technicality." This gives the concept an astonishing legal status, revealing its fundamental protective quality and ability to offer a form of preservation against abuse and violation. The "sheltering" feature of a dwelling or household or of an ecological space that offers sustenance manifests itself in the concept's source of power on the judicial register. The inhabitant is "directly connected to a habitation," and this connection cannot be violated, even after years of absence. *Black's Law Dictionary* (the definitive legal reference, cited often by the Supreme Court) emphasizes the term's difference from the more generalized and less emphatic idea of residence: "The terms 'resident' and 'inhabitant' have . . . been held not synonymous, the latter implying a more fixed and permanent abode than the former, and importing privileges and duties to which a mere resident would not be subject."[124] By 1866, the appearance of the term in cases clearly indicates its weakened, post–Gateward's Case status: "A Grant to or Prescription by the 'Inhabitants' of a place is too vague and not good; unless the Grant be by the Crown and then it erects the Inhabitants into a Corporation (*Willingale vs. Maitland*)."[125]

Inhabitancy is a concept of the human indicative of a design that precipitates the ideological constitution of the subject as the Cartesian, or "self-present," self (i.e., the metaphysical or "autonomous self" assumed to be endowed with an innate and unchanging substance that transcends the contexts of historical, geographical, national, and political occasions). It shares these qualities with the postmodern "decentered self," yet it differs in terms of its indissoluble relation to an ecological set of relations. It is not a form of human subjectivity designed to function as an instrument for the production of high yields. Nor does it initially appear as the formation of a subject geared toward the direction of its existence for purposes of national sovereignty (modern or otherwise). It is a conception of the human that speaks to a different understanding of economics and justice. The idea is based first and foremost on a cohabitation with the land, not on the desire to aggressively challenge the land for profit and energy consumption. Neither does its essential relation to *ratione commorantiae*—the act of "dwelling-staying-lingering" that arises out of the open access character of the common—have its telos in a blood-and-soil politics. Inhabitancy's connection to the right to a nonaccumulative (noncapitalist) sustaining of life is based more closely on what might be called, to use the current vernacular, "food sovereignty" and "habitation sovereignty"—a form of performance designed to oppose, in today's world, international actions like the Doha Agenda (2006), privatization, and the general militarization of the environment. It names an indissoluble, nonexploitative connection between the land and its human and nonhuman inhabitants.

It is not designed to take up a position of national security, ethnic security, or individual security in relation to the land. It sanctions a different relation to "universality," one defined not by a positivist phenomenology but by a postenclosure, postmetaphysical understanding of the universal as an "empty space" (to speak like Ernesto Laclau) not ruled, to invoke Black's legal definition, by a legal "technicality."[126] Such an emptiness allows for the particularities of habitations to exist, made possible in part by the heterogeneity and location-specific paradigm of custom law. As I will turn to examine now, it was precisely this kind of formulation of inhabitancy that the Brazilian dictatorship, and the neoliberal corporate state that replaced the dictatorship in the 1980s (both of which were supported by the U.S. national security state), attempted to erase during its developmental endeavors. And it is precisely this kind of formulation that the MST successfully revived. In contrast to the official security state memory of Brazil offered by Zubrin and others, an exploration of the counterhegemonic practices of the MST can offer a productive genealogical alternative and a vital countermemory to the dominant representational apparatus of enclosure-informed environmentality.

Though it is only one of the many and growing number of landless and communal movements operating as alternatives to contemporary globalization, the MST is of special interest for several reasons. It is the largest, and perhaps most successful, social agricultural organization in the world.[127] Founded in 1984 by a small group of squatters, its practice of occupying *latifundios* (large estates) expanded from 119 occupations a year in 1990 to 505 a year in 1999, with a total of 2,210 occupations in ten years.[128] The movement is said to have more than 500,000 occupations today (comprising approximately two million people) in some three thousand legally owned settlements.[129] These settlements have been able to develop an expanding architecture, organizing farm co-ops that number more than five hundred. These in turn support productive forms of marketing, credit, and technical assistance. Despite the "organic," heterogeneous nature of the settlements, the MST has been able to maintain and expand its distinctiveness as a large organizational structure that has influenced funding at the state level to "support its administration of 1,800 elementary schools with more than 160,000 students, teaching basic literacy to 30,000 teenagers and adults."

A brief examination of the MST's organizational structure illuminates an ontological understanding of existence that is based on a complex, singular connection to the land that stands against the distanced, panoptic perspective that abuses the land and its people for the gain of private corporations and the economy of unequal resource distribution that defines neoliberal globalization. It organizes its complex structure around a series of preventative mechanisms that are designed to ward off the mechanisms of widespread regulatory governmentality that were coming into existence during the first centuries of England's enclosure movement and now define twenty-first-century global relations. The idea of immanent collective leadership, for instance, is of paramount importance to the MST and directly opposes the kind of globalized legality that served as the self-justifying ground of common law. Experienced MST spokespersons offer their knowledge of past confrontations and negotiations with corporations and the government but refuse to take over the leadership of new camps arising in different locations. Experienced members of the movement believe that only those living in a camp should speak for that camp.[130] The movement thus wards off the formation of what we might call the panoptic figure, the transnational individual who, under typical circumstances, would assume a role as a primary spokesperson for an activist organization.

This prevention mechanism is internal to the organization's structure and carries great importance. From its origins, the movement has included as part of its basic structural apparatus the critique of the professional speaker-actor: "The people camped at Encruzilhada Natalino—and eventually the whole landless movement—came to agree that attacking authoritarianism by creating a new group of professional spokespersons to challenge existing authority would be unsuccessful, for they would themselves become authoritarian and paternalistic in the process of representing the group" (66). This critical awareness prevents the rise of a special intellectual class that would stand in opposition to a working class and enables farmers and workers and peasant families to be valued on the same level as priests, agronomists, teachers, lawyers, and doctors. The MST understands the importance of warding off modes of representation based on a metaphysical logic of a single, privileged group speaking for all. The preventative mechanism enables the group to function along the lines of a more genuine mass-movement model rather than a vanguard or trade union model (75). The prevention of one person speaking for another is also used to address the common problem within labor movements of unequal gender relations. Recognizing a lengthy history of gender inequality, the movement emphasizes the need for the public empower-ment of women—not only as active participants in the political arena, but also in the private world of the home, where gender lines tend to remain traditional: "The MST has found, therefore, that most of its efforts are aimed at nothing less than creating a 'new man and a new woman for a new society.'"[131] However, despite coordinated efforts to implement these mechanisms, the movement has not always succeeded in making them fully actualized. Efforts to break down the gender divide, for instance, were not always successful in all the camps, and because of this, a number of women have left the MST to form their own organizations.[132]

Even so, its prevention mechanisms are multidimensional; the move-ment attempts to address each of the many points of contact through which the security society and globalization attempt to influence the lives of those in agricultural communities. Monocropping—the favored form of production by agribusiness and one of the chief reasons for starvation among poor farming communities in Brazil and elsewhere—is one of the main forms of exploitation the movement confronts. Monocropping is a technique of imposing a single crop yearly on a plot of land without any rotation. It is the ultimate legacy of the enclosure movement's push for high yields, for it is considered to be economically very efficient and highly productive, producing great amounts of staple food and energy crops for the national and global markets. But its adverse effects are mani-fold, revealing the extent to which enclosure's high-yield philosophy has impacted today's global food and energy consumption. According to the Food and Agricultural Organization of the United Nations, the gen-eralized worldwide production of genetically uniform, high-yielding monocrops has led to the loss of 75 percent of plant diversity since the 1900s, and more than 90 percent of crop varieties "have disappeared from farmers' fields."[133] Such an extensive worldwide ecological transforma-tion is no doubt one of the greatest "accidents" of modern technological development, requiring, in turn, even greater enforcements of adaptation. By 1999, more than 75 percent of the world's food was being "generated from only 12 plants and five animal species."[134] Only three plants—rice, corn, and wheat—make up 60 percent of the food consumed by humans. The nationalization of agriculture indicative of the enclosure movement and the food and textile (sheep) marketing system it supported has led to the globalization of these same monocrop systems. According to the UN, "The extension of industrial patenting, and other intellectual property systems, to living organisms has led to the widespread cultivation and rearing of few varieties and breeds. This results in a more uniform, less diverse . . . global market."[135] As a consequence, small-scale farmers and the production of diverse crops and breeds of domestic animals have been heavily marginalized.

Monocropping, like enclosure, imposes a rigid, unidirectional pattern of development on an ecosystem, a pattern that needs to enact greater methods of security in order to maintain its existence. It is part and parcel of the asphyxiating, self-destructive drive of the Accidental ontology that I detailed in the introduction. Its Accidental character stems from the imposition of a closed system that self-generates greater and greater threats to its existence, thereby seeking and demanding greater and greater forms of security. On a basic level, the land is consumed by a single crop, or "identity," meant to transcend any other living entity. But this "basic level" should not be taken as a simple, self-evident answer to the problem of global demand for food, as monocropping advocates argue. It constitutes the security society's historically familiar pattern of *practice in crisis*—that is, the practice of imposing the presumably most efficient solution, the one transcendent crop guaranteed to produce the ultimate goal of the highest yield, which thereby forcibly reduces the heterogeneity and complex interactions of an ecosystem to a single, anthropocentric principle of being. The imbalance that this destruction of diversity generates consequently requires additional forms of manipulation that, in turn, generate additional insecurities.

The imposition of the "one" solution therefore gives rise to a cascade effect of "accidents." Increasing measures of "biological warfare" are necessary to ensure the continuation of the crop. The repeated action of planting the same crop in the same soil compromises soil ecology, depleting soil nutrients, which in turn removes the soil's buffers against parasitic species and increases the crop's vulnerability to insects, microorganisms, and other plants. Pesticides and chemical fertilizers are introduced into the environment to combat this precipitous ecological imbalance. These in turn adversely affect the surrounding ecosystem, in addition to the monocropped land. (Sugarcane and corn production—Brazil's major crops for producing ethanol for the energy industry—are highly dependent on ecologically destructive chemicals.) The crop, which is of a very narrow genetic base, itself "requires" genetic engineering to grow accustomed to greater forms of insecurity.[136] More engineering is required to protect the crop from changing climactic conditions. Specialized farm equipment is brought in to properly and efficiently plant and harvest the crop. Too expensive for most communities, the equipment needs to be financed, which in turn creates an economic imbalance for communities having a tradition of maintaining economic self-sufficiency. Communities and individual farmers are forced to take out loans, which introduce spiraling debt. This imbalance often subsequently becomes a political imbalance when agricultural production and distribution extends across national boundaries, and the tensions that arise from the lack of resources and finances are redirected toward the political neighbor next door (who is cast in the role of the "threatening Other causing all our problems"). Industrial harvesting equipment, specially designed to handle the cash crop, cannot process different varieties of other crops and ends up being useless for communities needing harvesting equipment for a diverse production of crops.[137] In addition, local communities historically constituted in a habitation that supports a diverse form of agricultural production often have many different uses for a single crop: "Rice is not just grain, it provides straw for thatching and mat-making, fodder for livestock, bran for fish ponds, husk for fuel. Local varieties of crops are selected to satisfy these multiple uses."[138] Genetically engineered high-yielding varieties (referred to as HYV crops) are designed solely for their grain, which eliminates a plant's multiple uses and forces the community to search for alternatives that then impact the ecosystem even further.

Brazil has a long history of monocropping that extends from the days of the colonial expansion of enclosure to the present. The Arab oil crisis of 1973 exacerbated the process of monocropping as Brazil turned to a nationwide production of ethanol to eradicate dependency on foreign oil. In the "Proálcool" program, the military dictatorship enclosed massive amounts of land for sugarcane plantations, offering large subsidies and a variety of incentives to individual farmers. Brazil's economic policy continues to be based on monocropping, or "agroenergy," for export and works in conjunction with large, foreign corporations and landholdings. The system focalizes energy and food production for national and international consumption and energy security.[139] The government guarantees the European Union, Japan, and the United States access to these resources and encourages countries in the Global South to adopt this agroenergy production model "through technology transfer."[140] The United States underwent a similar process in the early twentieth century, when the majority of small farms were consolidated into large farms and eventually brought within the hands of large agribusiness firms. This engrossment was seen as an advancement over former farming diversity, for it generated new technological forms of modern farming mechanization, advanced chemical growth methods, and improved water management, in addition to higher yields.[141]

Dario Paulineli, secretary of agriculture and environment in the Brazilian municipality of Luz, details the cascade of adverse effects: "Sugarcane has expanded rapidly in the last few years. Companies make contracts with farmers to rent their land, and the environmental impact is enormous. They apply toxins from planes which affects the populations of the cities. They destroy protected species of trees, plant sugarcane next to natural springs which feed the rivers, and they do not respect environmental impact studies. Many animals are dying with the devastation of the vegetation."[142] Farmers are forced to work for large, transnational corporations, and these corporations, in order to produce the highest yields, target the richest agricultural lands in the country. The cash crop supersedes all other needs, and crops for local food needs are often expressly forbidden. When not grown on local farmland, food must be purchased at groceries in town. When a downward fluctuation occurs in the economic market, farmers lose control over their ability to maintain livable levels of calorie intake. Depending on food sources from off site (at the market, which more often than not is stocked with imported food) means having to submit oneself to the unknowable fluctuations of the global economy. If global food prices increase, such a change can mean instant starvation.

In some cases, farmers have been so subsumed by this history of producing a single crop that they entirely lack the resources and even the knowledge to produce alternative crops for local food needs. The lack of information is compounded when farmers must confront problematic introductions of genetically manipulated plants and seeds by corporations seeking to increase levels of crop production. The drive to create transgenic seeds came about in part to solve the "problem" of having to use more pesticides to handle the cash crop's imbalancing of the predator insect population. Genetically modified seeds are made to be more tolerant to pesticides, enabling more chemicals to be used. Some, for instance, are engineered with toxins derived from a bacterium that kills pests like caterpillars. However, pesticide-tolerant seeds only promote the use of more pesticides, which destroy the soil.

Despite a variety of practices and despite disagreements on how to organize, throughout the 1980s and 1990s, the movement in general maintained a strong ecological commitment to avoid genetic and chemical-dependent practices. In the early 1980s—when the movement was still in its infancy and had yet to blossom into a recognizable phenomenon, even after winning its first battles for land—many of the farmers slipped back into the use of agrochemicals that pollute water sources such as rivers, ponds, and lakes. Roundup was a popular pesticide, used frequently and habitually. However, many members of the movement quickly decided that they needed to engage in more sustainable forms of agroecology. Agronomists working on behalf of the MST explained to farmers working in different locations that all ecosystems develop resistance to bacteria introduced into the environment so that eventually the bacterial toxins of any pest control will "almost certainly" lose its efficacy.[143] The movement gradually dispensed with artificial chemicals, turned to developing forms of fertilizers from animal and plant waste, began to compost, engaged in rotational grazing systems, and started to offer training courses in its affiliated schools in agronomy (80–81, 94). These changes eventually spread to the movement as a whole, though a number refused to change old habits. Making this break from old habits is difficult. Such ecologically minded changes are often thought to be either too time consuming or too expensive. Chemicals and transgenic crops are considered to be smart, scientifically advanced forms of improving one's production. But as Wright and Wolford discover in their conversations with MST members, eco-friendly agricultural land reforms ultimately provide a more secure safety net for the local farmer. "Success" is measured in diametrically opposing ways in the traditional, "counted" conflict between the farmer and the corporate entity. For the corporation, success means high production and profits from the land. This can only damage the land, as the agricultural "green revolution," which pushed unrelentingly for HYV crops in the 1950s and 1960s, starkly revealed.[144] For the small farmer, being able to engage in forms of agroecology and produce a variety of foods for daily needs means survival, even on a low level of cash income. Having to pay for expensive chemicals and chemically based crops that cannot thrive without the use of such artificial stimulants further exacerbates this situation. Most chemical practices of fertilizing are expensive and based on the large farming system; small farmers and collectives cannot afford such technological "improvements" and consequently suffer more when they are forced to expend funds on their purchase. When a downturn occurs in the market, the small farmer can lose everything overnight and in the morning find himself or herself subsumed by a large landowner who is able to survive such fluctuations.

Ensuring the quality and life of the earth is not a side issue for the MST. For the landless, the health of the land is paramount. The land provides the basis from which one can struggle for freedom. This concern obviously demands ecological knowledge, but it also requires a "whole systems" understanding of the many threats to contemporary existence: "A . . . fundamental commitment of the MST was to the earth itself. The MST understood that farm families could not find security on deteriorating land, a problem all too familiar to Brazil's southern farmers, who had seen the devastating consequences of sharecropping, land renting and absentee landlords."[145] The "commitment" is expressed here precisely in terms of resisting the kinds of land reform practices espoused by the discourse of enclosure. The MST understands that this concern for the human inhabitants of the land, however, must also express itself in a concern for the land itself: "In a meeting of the movement's rational congress, delegates from the local and regional organizations had approved a very strong statement of commitment to environmental protection. Chemical-dependent soybean and corn production was, as everyone who spoke at the meeting agreed, completely inconsistent with the conservation of soils and wildlife. . . . 'You have to look at the whole system in its natural context and design with that in mind'" (94). Moreover, this concern for the land manifests itself in the singularity of the local absolute: "There is no one recipe for how to do this, said the agronomist, which is why each farm needs to develop its own plan appropriate to its place" (95). Setting these protective mechanisms in place is not always easy for the MST. Some members have long been accustomed to counteragroecology practices favored by large corporations, which operate according to the speed paradigm of rapid crop production and the rewards of high yields, no matter what the eventual cost to the environment. These methods sometimes produce immediate gain in the short term but always damage the land and the prospects of the farmer in the long term. The MST takes a hard line on transgenic crops and announces at meetings that it will confiscate such crops if they are found, much to the dismay of some of its members. Importantly, this is not only the stance of the MST; transgenic crops are illegal in Brazil, and many nations in Europe have also banned them for similar ecologically concerned reasons. Such positions are not popular with the U.S. market, which pushes scientists to generate new and improved forms of seeds each year. Despite all these concerns, use of the transgenic seeds means that a corporation located metaphysically outside the local absolute of the land—an entity, in other words, that has no relationship to the land other than that of abuse—will ultimately own and control the land, the farmer, and the crops they produce.

The problem of lacking knowledge also highlights the way in which modern agribusiness and its emphasis on a form of agricultural production that requires increasing methods of control has begun to generate a new and privileged system of knowledge production. Chemical advances in pesticides, biological experiments in transgenic seed production, and engineering developments in farm machinery all take place outside local farming communities. The knowledge being produced in corporate labs and universities is of an increasingly specialist kind and is not available to the workers of the Global South. Moreover, this information (genetic information encased in seeds that farmers must buy in order to survive) is legally controlled by corporations that have patented this information. As Michael Hardt and Antonio Negri point out in their analysis of the change from Fordism to the post-Fordist era of precarious labor relations, "The process of agricultural change and the struggle over rights are increasingly dependent on the control and production of information, specifically plant genetic information."[146]

Thus, over a thirty-year period, the MST has struggled to find a variety of preventative alternatives to monocropping and its cascading, accidental effects. The MST attempts to delink the growing dependency on products and knowledge generated by the Global North and West, promoting an agroecological understanding of what it means to be an inhabitant of an environment by safeguarding the interactive and balanced relations active in an environment. Members of the organization focalize the need for diverse crop production; they encourage farmers brought up within the monocropping system to eliminate pesticides and replace them with organic fertilizers; they attempt to restructure ecological relations so that synthetic chemicals and modified seeds need not be purchased; they combat the loss of biodiversity; and they emphasize the production of food for local consumption, in addition to the cash crop.[147] Hence the organization, despite setbacks and differences of opinion, has nonetheless promoted a deep and critical awareness of ecological habitations and a deep and critical understanding of the many forms and practices of inhabitancy.

Indigenous Movements: Inhabitancy and the Redefining of Politics

We can already see that the MST is a productive, nationwide organization that successfully opposes neoliberal policies of enclosure and development and the neomilitary structures of security. A member of Via Campesina (the international peasant farmers organization—the organization that first coined the phrase "food sovereignty"), the MST is considered to be "the largest and most politically coherent national movement within Via Campesina."[148] Despite its size and strong organizational presence, its leader João Pedro Stédile nonetheless maintains a healthy critical awareness of the organization's progress and power, arguing recently that the MST's effectiveness in organizing the poor is still limited.[149]

We can begin to think how inhabitancy as a conceptual term describes the organizational and relational actions of MST members. There are many ways the concept and modes of preenclosure habitation can be found operative in the organization. The MST marks its identity, as I mentioned earlier, in relation to the counterhegemonic terms "food sovereignty" and "habitation sovereignty." Food sovereignty names the need for a people to constitute and define their own food systems in terms of production, consumption, and distribution.[150] It defines food as a basic human right, advocates for an end to the globalization of hunger, and pushes for an agrarian reform that also attends to the protection of natural resources. Habitation sovereignty names the fundamental human right to a habitation and a habitable environment. Both terms are closely related to the preenclosure custom of "use rights." The term *sovereignty* needs some explanation, especially given its negative history in poststructuralist theoretical critiques of "sovereign rulers" (of domestic and foreign territories) and "sovereign individuals" (a subjectivity produced by the ontology of the Cartesian metaphysics of presence, the discourse of enclosure, and the economy of capitalism). Sovereignty in the sense used as a countermeasure to traditional security and capitalist local–national–global conflicts names a performative action of resistance. It should not be confused with a Kantian idea of the subject or a Husserlian phenomenology; rather, it is more closely related to a form of nomadic action that

Deleuze and Guattari have described as "holding space."[151] The problem with so many of the postmodern ideas of subjectivity—fluid, socially constructed, contextual, and so on—is that they fail to fully consider the need to couple the concept of decentering to the equally important concept of "radical exteriority" that decentering opens up. All too frequently in these theorizations of postmodern subjectivity, exteriority is presented as an abstraction. Against these theoretical abstractions, we should posit the following: there can be no concept of exteriority that is devoid of place. Exteriority is always a matter of a specific situation—that is, of an environment. To put this differently, there is no decentered subject without an active engagement with a territory. More often than not, this engagement with a situation manifests itself as a struggle, for in the neoliberal and security societies of the twenty-first century, all places are viewed simultaneously through a lens that sees the world's places in terms of enclosures and in terms of a security risk. For this reason, I find the term *situational* to be more productive than the more abstract *exteriority*—for situations are always about the specific needs of a place, a people, and an ecosystem of relations, and more often than not people are situated, and resituated, against their will.

In this sense, the MST cannot be understood as being committed to the rights of people in some environmentally dismissive, anthropocentric fashion, which in many cases describes the nature of antineoliberal, socialist-oriented movements. To avoid such anthropocentrism, the MST carefully articulates its commitment to ecologically sound agricultural practices. The movement's foregrounding of the need to find green agricultural practices provides us with a strong example of the way in which their theories of subjectivity speak to a specific situation and not just some abstract notion of a postmodern exteriority. For instance, practicing sound agroecological farming means that the movement must learn about the specifics of place: "The underlying idea of this model is to use the knowledge of ecological processes to design agricultural production techniques to enhance the health and diversity of the landscape consistent with agricultural productivity. There is no one recipe for how to do this [says one of the movement's agronomists], which is why each farm needs to develop its own plan appropriate to its place."[152]

The MST also actualizes the concept of situation by employing a number of warding-off strategies that take a firm position on transgenic crops. As Vandana Shiva has powerfully argued, transgenic seeds threaten both the food sovereignty and the habitation sovereignty of small farmers.[153] The MST realizes that patented seeds benefit an entity metaphysically distant from a local habitation. The plant that produces the food will always be owned by someone else—not by the indigenous population. It breaks down the collective, relational, and situational structure of the habitation—both in terms of people and in terms of the environment. Neoliberal governments and corporate entities classify such refusals as "protective mechanisms" that thwart efforts to introduce sustainable development (a particular deployment of "sustainable" by the free market). José Placotnik, one of the veterans of the MST, articulates how his philosophy of resistance depends on ecological connections: "A . . . quail flew up in front of us and [Placotnik] explained that while they were good to eat, he didn't hunt them. 'The whole point of keeping the woods in place is to leave places for wild things, which have their own place in the world and should be valued as humans are valued.'"[154] For Placotnik, this philosophy of "letting be" forms the basis of a daily existence orchestrated toward the sheltering of habitations from violation.

Throughout its historical development, the MST faced numerous military aggressions, either in the form of private militias working for large landowners, the national military, or the pressure from defense departments outside the nation. The inviolable enclosure rights of private property were under threat, and large landowners, rural business leaders, and national organizations characterized the MST and its agrarian reforms as a threat to peace and the stability of the nation. Key to the MST's success against these various forms of security society, military actions was the concept of holding space, of squatting on the land in an act of signifying a nonessentialist commonality—the "dwelling-staying-lingering" operative in the inhabitant's claim to *commorantiae*—that was utterly denied by the powers that be.

This simple act of refusing to move is the very heart of an ethical politics that enables what we can characterize as a universalizing praxis, but a praxis that does not colonize the space of the universal with an identity that would need to securitize its existence. The essence of the very act of this resistant immobility puts into play a form of effective commonality based on what is not allowed to exist in the sociopolitical field of reality, the governing "arithmetical play of profits and debts," to invoke Rancière.[155] As Placotnik argues, "The point is we're trying to change the whole concept of politics, the relationship people have to the politicians.

Not to set one up high and the other down low. You have to think about the meaning of the whole style."[156] Wright and Wolford also articulate the failure of a modern legal system to address this kind of affective politics of inhabitancy. They invoke the overlooked importance of Fernando Sodero, one of Brazil's scholars of land law who worked in the latter half of the twentieth century: "[Sodero] came to the conclusion that legalistic interpretations of the land law were futile. . . . As a legal scholar, Sodero could find no just solution in the letter of the law itself and proposed that only enlightened judges using a general philosophical sense of fairness, along with meticulous and profound examination of the historical and human circumstances of each case, could reach reasonable settlements of land claims. The law would be of little help."[157] The act of squatting performed by the MST constitutes the establishment of a political bloc that is based neither on some form of metaphysical commonality (such as the kind of fundamentalist communalism that leads to ethnic cleansing) nor on the dynamic of populism (such as that of the Tea Party Movement in the United States). Rather it maintains its community as that of the landless.

The very word "landless" itself names a fundamental ontological absence, an inequality unlike the inequalities claimed by other forms of political movements. The inequality that lies at the heart of the landless is a contentless absence of equality that exists prior to all measureable forms of equivalency and inequivalency. It cannot be said to be an inequivalency that has meaning within the system, for it cannot even be measured by the system; its existence as an inequality is not of a measure that the system recognizes. The global capitalist system of development understands inequalities very well; it cannot exist without understanding them. The various levels of inequality are identified and thoroughly calculated within its structure: from the executive, to the local emissary, to the workers in the assembly lines, to the managers of various districts, to the large and small farmers, down to the migrant and day laborers that work for large and small farmers. The squatting inhabitant that *holds the space of land* is an entity that has no position in this calculated system of inequality. As such, it speaks of the very *idea* of inequality—not the inequality of a particular group as it is measured in relation to another group already identified by the system. These already identified groups are forced—by the very presence and structure of the system itself—to measure their lack in relation to the instituted and planned hierarchy of those above and those below. This derivative thought of inequality manifests itself in the positive terms of an "us" versus "them" economy. It is an idea of inequality generated by the system itself and is one of the means by which the system convinces its subjects to struggle against one another. In other words, this reductive thought of inequality offers no avenue for thinking an end to the system of inequivalences; it is merely an idea designed to perpetuate domination and the rule of capital accumulation. At best, it generates the pathetic performance of philanthropy, which ultimately does nothing to end the inherently repressive nature of developmental projects.

Inhabiting a space of land through occupation, however, brings to the foreground *a form of inequality that applies to all*—the concept of inequality itself, not one that is measured and identified as the inequality specific to a particular group or identity. To mark this particular formulation of inequality, I will use the phrase *the form of inequality*. This "form" is not part of the system; it has no position in the system (squatting is not recognized as a viable activity or standing for an identifiable subject). It is a form of contention not interpellated into the system (contentious struggles to make oneself less unequal than others by climbing the corporate ladder, for instance, are recognized as clearly identifiable and positive forms of taking action—the very heart of being a productive subject in the logical economy of neoliberalism). The MST as a community of squatters exemplifies the "part of no part." They are the people that have been "propelled . . . into . . . nonexistence" by what the system wishes to hide: the fact that the measurable forms of inequality all lead spiraling down into the ultimate immeasurable emptiness of the nothing-at-all. From within the space of this nonposition of the nothing-at-all, there is no point from which one can measure one's inequality. One comes together with others to name the wrong that is the form of inequality.

It is important to remember that Rancière's ontological theorization of the political, despite the limitations of its tight disciplinary philosophical focus, still holds great importance when we begin to look, for instance, at the increasing inequalities that mark the distribution of wealth. The MST arises in reaction to such inequalities, and its unwanted presence in the national and global scene revisualizes such distribution as the form of inequality itself. This crucial nonpositive element of "having nothing at all" is written into the very name of the MST: *sem-terra* means literally "without land." But this "not having land" should not be thought about from within the traditional discourse of colonization—that is, the point is not to acquire a piece of land so that one can finally join the legions of those profiting from their privatization of the ecosystem. The relation the MST has to land is grounded in the activation of the concept of political freedom, of naming the wrong of the miscount that is inherent to the environmentality of neoliberal development. This naming of the wrong essential to the enacting of the political is realized in the coupling of a protective understanding of the land as a foundation for existence, a foundation that is an explicit right not of any one particular faction but of *all*. This coupling of land as a foundation and a fundamental right is engendered in the political act of occupation: "One MST activist . . . says the movement was successful in expanding across the country because 'we picked an issue that united everyone—the land. The land is the necessity, it is the call that unifies everyone. . . . You offer the workers the opportunity to have land—but [you introduce the idea] of a mass occupation.'"[158]

What I have been calling the form of inequality is similar to Rancière's references to what he calls "the fundamental miscount of politics," or "democracy's miscount."[159] Rancière's formulation of the "miscount" makes apparent the violence that undergirds policing politics—the realization that any political regime or formal order is founded on some incorporation of unequal positions, an isomorphic accumulation of difference that it "counts" in order to establish a recognizable and calculable structure. This subsequently enables the formation of administrative methods for securing the reproduction of this structure. There is always some "miscount" in a system, some nonexistent possibility that is not counted by the system on principle. In the discourse of environmentality, that miscount is inhabitancy. The strongest aspect of Rancière's theorization of the miscount of neoliberal democracy is his attachment of the event and struggle for freedom to the part of the population that is not counted. All the subject positions within the system that are counted, from the top to the bottom, do not contribute to the potential for freedom; they are designed precisely to maintain the system and, as such, work against freedom. For Rancière, freedom is an agential quality, or activity, that is peculiar to those who have no part—those who have nothing and who express an activity and struggle that is not part of the system. Freedom is thus a name for the ontological demand for a fundamental interruption of the order of domination. It presents an act of decoupling from the movement of neoliberalism's speed, acts of security, and the structure of the Accidental. It stops the current of speed, and at the same time—because it lies outside the materiality of the system, the substance of the system—it is not an accident produced by the system. The nonparty of holding land in the MST is of a magnitude entirely different from the appearance of the accident, for the accident is an *integral* part of the security paradigm. The dwelling-staying-lingering of inhabitancy is not a momentary stopping of the current/currency of the Accidental and its speed; it "pulls the plug" on this terrorizing current.[160]

The MST practice of only letting local members speak for a camp is a perfect example of a form of political activism that productively avoids global policies of what Virilio refers to as "speed and politics." This Freire-influenced practice not only avoids the problem of paternalism but also speaks directly to the power of maintaining what Deleuze and Guattari once called a "local absolute": "an absolute that is manifested locally, and engendered in a series of local operations of varying orientations."[161] The local is "absolute" in the sense of an action that applies only to a specific location, as opposed to a "relative global"—an action that depends on a generalized, global economy that limits, defines, and, in the final instance, supersedes the authority and power of any specific situation. The global peacekeeping orders and their concern for "victims" is one way the security society attempts to take away the power of the local absolute from a group of inhabitants, for victims cannot act for themselves and are always in the position of relying on the power of others, who typically use victims as pawns in some larger geopolitical strategy game: victims, put differently, are the accidents of imposed political regime. The MST does its best to maintain its character as a local absolute organization even in the face of sympathetic organizations such as the Workers' Party, the church, and unions. In the early 1980s, when the Workers' Party (under the direction of Lula, later to be Brazil's president) had helped with several of the MST's land occupations, the majority of the MST delegates still decided—when they had the chance to let the more powerful Workers' Party take over—that it would always be best for the MST to maintain its independence, even when seeking out ways to form productive alliances.

Coda: The Right to Inhabitancy

The emergence of counterhegemonic groups like the MST reveals the continuing (and growing) presence of the discourse of enclosure and the monomaniacal drive to control production on a macroplanetary and a microbiological scale. As I will argue in the chapters that follow, the present age—in the context of an environmentality that functions by foreclosing all but the empirical, which makes practically impossible the genuine political act that breaks through this reigning ideology—should more rightly be reperiodized as the age of the global war against inhabitancy and the fulfillment of the enclosure movement that began centuries ago. This war bypasses the needs of the environment—and the very presence of nature itself—in favor of a version of science based on the technological "advancements" of speed, efficiency, and high production, advancements that ultimately result in greater ecological accidents and strengthening forms of securitized adaptation. As during the eighteenth-century enclosure movement, the twenty-first century's "energy security" discourse rewrites nature as a recalcitrant force that must be ruled at all costs. The laborers and peasant movements not associated with a privatized international mechanism are a threat to these modern-day feudal lords of the land. Growing "technologies of development" arise out of this history. But as we've already begun to see, they are also developed in relation to a historical process of generalized military expansion that began in the eighteenth and nineteenth centuries. In the next chapter, I will explore how this expansion sought out new forms of securing populations and their energy reserves by forming connections with both the formal and the natural sciences. As colonial forces of transnational control evolved into different formations between the years of World War I, World War II, and the Cold War, these connections became increasingly hardened and naturalized. Land development projects would soon become indissolubly related to technological developments in engineering, systems of surveillance, and computer processing. Such connections are difficult to trace, given the limits of current disciplinary relations. Yet they are the essence of today's accident-oriented environmentality.

Struggles like those engaged in by organizations like the MST reveal the power of the resistant force of inhabitancy, a conception of the human subject, as I've been articulating, that is fundamentally different from constitutions of subjectivity offered by globalization, nationality, and entrepreneurial individuality. National and international law, because of their foundation in the discourse of enclosure, cannot provide a resistant politics for the global poor involved in such organizing practices. The principle of individual sovereignty equally destroys the chances of the global poor to organize against large corporate and state landholders. What is needed, ultimately, is a new judicial conception and deployment of food sovereignty and habitation sovereignty, and the idea of inhabitancy speaks to both of these. Food sovereignty names immediately the pressing issue of planetary starvation. Habitation sovereignty names the indissoluble connection between the land and its human and nonhuman inhabitants.

Both of these, I submit, offer real potentials for countering the brutal forms of corporate and militarized feudalism. Early feudal enclosures and later parliamentary enclosures have returned in the guise of corporate and national securitizations of the planet. Such actions are especially manifest in the discourse and practices of energy security, with neoconservatives and liberals working in tandem to enclose the land on a planetary scale. In the political push for energy security in the consumer world and in the struggle for basic existence in the developing world, we can observe the evolution of two kinds of human subjects: those that seek to control the environment and those that are connected to (and brutally disconnected from) their environments. If ideas of energy security continue down the path of history's enclosures, environments and their inhabitants will be too compromised to save. The poorest communities across the planet already live with tenuous connections to their habitations. The transformation of corn and other edible crops into fuel for developed countries will inevitably cause prices in staple foods to fluctuate at rates heretofore unknown. Even the World Bank argues that large increases in food prices will "raise overall poverty in low income countries substantially."[162] In the war against habitations, human subjects without a right to inhabitancy will quickly become the shadow kingdom of the twenty-first century. They will take on the mantle of history's dispossessed—the farmers, laborers, villagers, and tenants who lost their rights to inhabitancy during the great waves of enclosure acts in the eighteenth and nineteenth centuries. Is this not the essence of the great seventeenth- and eighteenth-century war against custom law and the enclosure movement's war against inhabitancy? Without this war, common law and enclosure would fall back into the nothingness from which they originated. The current historical ecological occasion would thus seem to present us with two very different questions: Should half the planet's population be reduced to an adaptive form of life that offers little opportunity for existence so that consumer society's high yields can be maintained? Or should the life of all be rethought from the standpoint of the right to inhabitancy?

3

Genealogies of Military Environmentality I

The Human Species as a Geological Force in the Anthropocene

The reason we did this job is because it is an organic necessity. If you are a scientist you cannot stop such a thing. If you are a scientist you believe that it is good to find out how the world works; that it is good to find out what the realities are.

—J. Robert Oppenheimer, speaking about the construction of the atomic bomb

Suddenly and without any sound, the hills were bathed in brilliant light, as if somebody had turned the sun on with a switch.

—Otto Frisch, eyewitness account of the "Trinity" nuclear test, July 1945

Journey of Death

Otto Frisch was one of two important German physicists who fled to England during the early years of World War II. In 1940, along with Rudolf Peierls, he produced the first theoretical paper confirming the possibility of building an atomic weapon. Less than five years later, Frisch witnessed the first successful detonation of the atomic bomb in the Jornada del Muerto desert in New Mexico. Exactly three weeks after that, the atomic device known as "Little Man" was dropped on Hiroshima. Approximately seventy thousand to eighty thousand people were immediately killed from the blast. Those close to the detonation point evaporated instantly. More than fifty-five thousand later died from the effects of radiation sickness. The bomb destroyed some six square miles of the city. The heat of the explosion caused fires to erupt across the city, lasting three days and killing many of the remaining survivors. Long-term effects of the bomb's radiation affected the living and their children for the remainder of their lives. The bomb was declared extremely inefficient: only 1.38 percent of its uranium ignited.[1]

Jornada del Muerto is Spanish for "Journey of Death" (or "Dead Man's Journey"). The first technological "ground zero," it lies far away from any dense human population. Because of this, the experiments it allows can occur in an intellectually purified space, a location aesthetically, geographically, and politically sanitized from any realization of the horrors of atomic destruction and radiation. At Jornada del Muerto, Frisch would attempt to capture in language his feeling of awe in beholding the first successful testing of the bomb. For Frisch, the event signals the triumph of science over the inherent mysteries of the universe, a moment of enlightenment so wondrous that it can only be likened to capturing the power of the sun. His eyewitness characterization of humanity's unleashing of the most destructive force to date emerges in the differently oriented space of scientific enlightenment and representation—dispassionate and coldly calculative but equally enthusiastic about technological progress. The comparisons he uses in his description, mostly drawn from nature, sound even childlike at moments: in addition to the now-hackneyed description of the "mushroom cloud," he likens the physical appearance of the expanding atomic explosion to a "ball look[ing] somewhat like a raspberry," attached to a stem "looking . . . like the trunk of an elephant." The immense cloud that spread suddenly above the explosion looked to him "like a pool of spilt milk." When he feels the actual blast of the explosion arrive, it sounds to him not like thunder but, as if he had just come from watching an American Western, "like huge noisy wagons running around in the hills."[2] Frisch was seeing the accomplishment of a lifelong goal. His earlier writings and documented conversations with other physicists involved in the project of splitting the atom are laden with the banal discursive embellishments of a popular adventure novel, crescendoing even to the climactic moment of the final discovery, when the pieces to the puzzle of mass destruction come neatly together: "The charge of a uranium nucleus, we found, was indeed large enough . . . ready to divide itself at the slightest provocation. . . . Einstein's formula . . . [H]ere was the source for that energy; it all fitted [*sic*]!"[3] His aunt and colleague, physicist Lise Meitner (whom Einstein praised as the "German Marie Curie"), wrote to him when she first read the conclusions of his theoretical work: "These results, I realized, had opened up an entirely new scientific path. . . . The [splitting of the atom] is a truly beautiful result. . . . You now have a beautiful, wide field of work ahead of you."[4]

Frisch's experience—the historically orchestrated excitement of scientific exploration in a time of war—embodies the central concern of this chapter: the extension of notions of security, defense, and control and their apparent natural connection to an organic biological and ecological world into monumental realms of ultimately uncontrollable and destructive power. His triumphant mobilization of immense energy, I argue, marks the militarized evolution, in the age of the anthropocene, of the human species into a *geologic force*. The satisfaction and awe that characterizes his and other physicists' experience in the early days of the atomic age exemplify the paradoxical coupling of security and insecurity that I have been identifying as a component of environmentality's self-destructive nature. But this new weapon—described as "beautiful" in its orchestration—is equally "sublime" in its limitless ability to demolish any system of order or means of defense. Such a colossal power radically transforms the nature of human influence. It changes the ecological status of humanity as a species. In the wake of this military mobilization of the atom—a nightmarish version of "energy security"—security pretensions will release greater potentials of insecurity on a scale previously impossible to imagine.

In his 2012 essay "Postcolonial Studies and the Challenge of Climate Change," Dipesh Chakrabarty argues that we need to expand our predominant images of the human to include an awareness of the human as a "geological force"—that is, "the figure of the human in the age of the Anthropocene, the era when humans act as a geological force on the planet, changing its climate for millennia to come."[5] Paul Crutzen's recent popularization of Eugene Stoermer's designator "anthropocene" serves as an effective identifier for articulating the scale of human impact on the planet's ecosystem. The term's significance lies in its provocative power to reveal a new image of the human, one that stretches our limited, proneoliberal imaging of human beings as products of Western Enlightenment individualism and industrial-era entrepreneurship. Both of these images of the human—as an individual and an entrepreneur—owe a good deal of their existence, as I argued in chapter 2, to the discourse of enclosure and common law and thus to the erasure of an image of the human as an inhabitant connected to and influencing an ecosystem. We can now see that the erasure of inhabitancy as an idea of the human also obscures an awareness of the human as a global force. This concept of the human as a "geological force," I will demonstrate in this chapter, is central to any critical understanding of human–environmental relations today. One, it serves as an effective way to characterize the kind of anti-inhabitant human subjectivity idealized by the security society in the age of environmentality. Its existence is the ultimate realization of the enclosing individual—the being that seeks to rise above environmental restrictions by owning, manipulating, and improving vast geographies. Two, it signals the beginning of the deployment of insecurity as the founding gesture of security, or what amounts to the same thing: the establishment of insecurity as an integral component of existence—what I referred to in chapter 1 as the Accidental.

As we will see in this chapter, this twentieth-century period of environmentality, which I described earlier in terms of the four phases of the U.S. national security state beginning in 1940, owes its existence to new patterns of Western military development that arose during the nineteenth and early twentieth centuries. As common law, the discourse of enclosure (a "beautiful" domination of the earth that generates the sublime accident of global warming), and later the Industrial Revolution spread throughout Europe and America, military bodies began to adopt methods of discipline and organization that matched these sociopolitical formations. Environmentality, in other words, was evolving into a *generalized architecture* of a militarized human sociopolitical existence. A new "target-oriented" methodology began to take hold, one that operated both materially (in terms of weaponry and land enclosures) and intellectually (as a new adopted pattern of thought—an unstoppable scientific search for the truth and an insatiable desire for economic improvement). This methodological imperative saturated various sites of production, becoming in the twentieth century a highly theorized form of conceptualization referred to in the field of military theory as a *higher-order machine*. This theory of the higher-order machine combined the overwhelming logic of a unidirectional pattern of thought with the belief in an organic form of organization and rationality. The mixture of these two constitutes the intellectual basis for environmentality. Their combination informs such simultaneously captivating and strangulating ideas as the "scientific impulse" to adhere to an "organic necessity," no matter how catastrophic the results (Oppenheimer's rationale for completing the atomic bomb); the defensive-driven belief in "natural security" as the militant basis of biopolitical existence (Sagarin); and the idea of "adaptation" as the logical solution to (the war game of) climate change.

But in order for this generalized architecture and the organizational ideal of the higher-order machine to become a reality, the war machine needed to overcome certain human limitations. In the seventeenth and eighteenth centuries, the erasure of inhabitancy that made possible the privatizing entrepreneur also generated a new human instrumentality, one that radically transformed the ways in which human beings impacted their environments. The value of the human now lay in the ability to become a force that, in owning greater amounts of land and redirecting that land toward the market-driven obligation of a higher yield, could transcend its former immanent communal relations (recoded as limitations to personal economic advancement). Not only a king or a military leader, now (potentially) every single human being had the ability to secure and command large environments for purposes of optimal production. This act of territorial transcendence introduced a new network of power, one that enabled certain privileged human subjects to acquire great ecological influence, changing the meaning of the environment and redirecting the course of its development.

In this chapter, I trace the rise of this new version of the human, exploring the precise enhancements that made it possible for humans to acquire the immense power to become geological forces. The specific twentieth-century evolution of the human, as we will see, is made possible through the establishment of a new military imaginary, a powerful analytic that extends the ideological demand for security into new realms of research and development, creating an extensive network of militarized movements. The rise in this human formation involves the generalization of militarized existence into a number of fields of human inquiry through which the security society expanded its efforts to maintain global supremacy in the twentieth century: mathematics, communication, systems analysis, data analysis, computer processing, game theory, and management theory. The combination of these disciplinary developments culminated in the central military organizational model known as C3i: command, control, communication, and information. At an ontological and epistemological level, these new intellectual formations prioritized "explanation" at the expense of a critical understanding that would reveal their precise nature and inherent limitations. At a material level, the new desires for command and control required the creation of new mechanisms of mobilization that could successfully orchestrate the activities of "tens of thousands of people" (to quote Brigadier General Thomas F.

Farrell, second in command of the Manhattan Project) so that they could work together like a machine to achieve the "fullest fruition" of the security society's ideal of technological thought. I explore these developments specifically by reconsidering the work of theorists such as Manuel De Landa and Paul Virilio and scholars such as Chakrabarty and Rob Wilson. I extend their work by looking at military theorizations of the "command and control" (C3i) rubric and the search for forms of "higher-order" control, considering finally the global influence of these developments on the rise of new disciplines such as computerized processing and game theory.

The Prisoner's Dilemma: Ecological War Games

Military interests in human evolution and control have a history of looking to both the natural and the formal sciences for supplying new ideas on how to extend the might and reach of the war machine's influence. Explorations in the natural sciences for organic forms of order, a possible inherent ground of security, worked alongside similar endeavors in the formal sciences—namely, mathematics, logic, and computer science. Answers to the problem of how to "streamline" soldiers' different sensibilities on the battlefield into a single, organizational motor fluctuated between naturalist and formalist explications, as did the desire to establish an esprit de corps in the civilian sector of the nation. Some efforts, such as the search for new targeting technologies, developed primarily within the formalist camp (the mathematics of ballistics, for example). The space race, in addition to its need for physics and engineering (rockets and propulsion), introduced new interest in the formal sciences, reinvigorating advances in computer technology, systems analysis, war game theory, and so on. This unique emphasis on the formal sciences established a new, somewhat otherworldly field of battle, one that was more "virtual" than the early concerns for discovering methods of organization on the battlefield. Perhaps the most decisive formal analytic governing the security society during the Cold War was the war game scenario known as the "Prisoner's Dilemma." The Prisoner's Dilemma was an experiment in game theory undertaken at the RAND Corporation in 1950. It was an attempt to discover a logic-based "ultimate solution" to the problem of Cold War nuclear tensions between the United States and the USSR. The game suggested that after calculating all possible variables and outcomes, there was only one rational course of action for the American military: to generate as many nuclear arms as possible. The invention of mathematicians Merrill Flood and Melvin Dresher, the Prisoner's Dilemma was *the* scenario adopted as an inherently logical foundation for this buildup of arms.[6] Like the Center for a New American Security's (CNAS's) Climate Change War Game, the Prisoner's Dilemma attempts to hide its lack of an ontological guarantee in the metaphysics of seemingly obligatory wartime contingencies.

The narrative of the Prisoner's Dilemma follows a particular, unidirectional pattern, a pattern governed by the general methodological imperative of the security society linear analytic we've been examining. The scenario begins with two criminal associates being captured and taken prisoner. Even though the prisoners are safely incarcerated, the authorities face the problem of not having enough information to obtain convictions. Both prisoners refuse to talk, and they are consequently separated and interrogated through a kind of mathematical charm offensive. Each is separately offered a choice from three possible deals: (1) the prisoner can betray his or her partner, in which case he or she will go free, and the one betrayed will serve a life sentence; (2) both prisoners can betray one another, in which case they will both serve a medium sentence; (3) neither can accuse the other (they both remain silent), and each will receive a short sentence. The artificial, militarized symmetry of the game leads logicians to assume that each prisoner will end up betraying the other, since supposedly neither will want to take the chance of serving the life sentence—even though it is to the advantage of both to cooperate and remain silent. Logic—thought to be an entirely demystifying form of war game intelligence—assumes that the prisoners in this scenario will perform like good Cartesian, entrepreneurial subjects, always already dedicated to an economy of (self) return. This single fantasy of a mathematical neutrality was the basis for the nuclear arms buildup of the Cold War. While the best choice is for each prisoner to remain silent, it was assumed that it would ultimately be *irrational* for one person to take the risk of thinking the other would not "betray" the situation by continuing the buildup of arms. In this sense, the game, thought to be grounded in a dependable rationality, provides us with a perfect example of the accident—of insecurity—being the "substance of reality" (chapter 1): from the beginning to the end, the player is forced to resolve the problem of how to live by incorporating the least amount of insecurity. Indisputable mathematical probability forms the basis of this "rationality" that neither prisoner will want to risk a life sentence (and in the case of nuclear war, complete annihilation). The "best" or "safest" course of action is thus always betrayal—because no matter what the outcome, one could avoid the stiffest sentence of life. In this sense, the war between the United States and the Soviet Union could never "rationally" be one of cooperation but *must always be* one of suspicion and defensiveness, despite the fact that both would benefit from cooperation. Logical rationality thus dictated that both sides must build up their supply of arms, despite the fact that this increased the risk of planetary destruction. Any other course of action would have been irrational. From this logic arose the original planetary accident of worldwide radiation and the twentieth century's fourth ground zero in the Pacific Proving Grounds (Hiroshima, Nagasaki, and Jornada del Muerto, of course, being the first three).

The "necessity of betrayal" that defines the Prisoner's Dilemma is the ideal scenario of the security society. As a militarized exercise, the game's goal is to maintain the rigidity of command but to blanket that command in the formal sciences—in this particular case the axiomatic givens of mathematical calculation. Its telos of constant security lies in its very beginnings, in a phantasmatic scenario designed to arrive at only one conclusion. The end goal of constant security overrides all other decisions, even if that means giving up the potential to cooperate with others. We can now begin to see that the very idea of security as constituted in military matters today is *essentially* the image and logic of an axiomatic positivism evacuated of political content—the disavowed antagonism of an inherent groundlessness. The "solution" of the Prisoner's Dilemma is assumed to spring from the truths of formal science itself, the highest attainable level of intelligence.

The double disavowal that silently buttresses this neutrality, however, begins to unravel when we consider the unexamined assumption of the exterior entity brought into the game in order to make it function—the unexamined assumption of exteriority itself as always and only *an entity that will betray*. In other words, the enemy-subject is built directly into the structural economy of the game, and cooperation is "problematized" as a stumbling block to be ultimately deleted. In its essence, the game mirrors the logic of the Climate Change War Game, in which adaptation trumps alternative scenarios for ecological existence. Put differently, the security society uses the event of climate change to shackle any and all human relations with nature to the "security buildup" of adaptive arms against an enemy that cannot be trusted, an enemy that is everywhere (in the atmosphere, in the seas, etc.) and in the process of building up its destructive arsenal. In the case of climate change, that enemy is both other nations (that pose a threat to energy security, food and water resources, etc.) and the environment itself (now replete with greenhouse gases). In this sense, environmentality names an age when the very substance of existence itself—a planetary environment filled with methane and carbon dioxide—has become the accident, the integral force changing the matrices of existence. Cooperation in the Prisoner's Dilemma and alternative ecological relations in the Climate Change War Game are the Lacanian Real of security logic. No matter what one does, cooperation always comes at too high a cost—one is always "sentenced" from the start. Betrayal and mass destruction, far from being events to be avoided, become dividends. And privileging betrayal, like the privileging of the accident in postindustrial human consciousness, is the preferred option in a system that wishes to operate through the administration of public fear.

We can now begin to see the logic at the heart of such phantasmatic "theory of everything" scenarios and the way in which these logistics inform sociopolitical, economic, and military exercises such as the Climate Change War Game. Adaptation is an "easier" answer for the military, because it enables one to continue the enforcement of security measures. Once this "anxiety of betrayal" is telegraphed to the general public, dispassionate and apolitical mathematics can move to center stage without question. If the dispassionate formal sciences can be made to constitute themselves from the ground up in the direction of targeting inhabitants, then the services of those who might challenge politics as a form of policing (to recall Rancière) will no longer be required. In this world picture, the ungovernable and uncontrollable ecosystem is not only the foundation of existence but also the ultimate enemy combatant, and choosing to cooperate—to *sustain* the ecosystem—would amount to committing the ultimate naïveté of risking the worst-case scenario of a long sentence.

We can begin to see the full meaning and impact of the Prisoner's Dilemma scenario by looking at the complex context in which it was developed. Both Flood and Dresher were heavily involved in the boom of Cold War explorations in military and security intelligence. Today Flood is considered a pioneer in operations research and management. His work is exemplary as an eschatological indicator of the folding of economics, game theory, information management, computer programming, "activity analysis," operations research, and industrial engineering into the security society in the post–World War II occasion. This complex enfolding of various disciplines established impressive productive flows between the civilian and military sectors to such an extent that the connection between the two became seamless. Besides the codevelopment of the Prisoner's Dilemma scenario, Flood founded The Institute of Management Sciences (TIMS), served as president of the Operations Research Society of America (ORSA), and claimed to coin the term *software* in a 1946 War Department memo (to distinguish the "ballooning research cost items that could not be directly attributed to military hardware budgets").[7] He would provide one of the first directly connected efforts between mathematical study and military targeting in his 1944 study on the aerial bombing of Japan with the B-29 bomber.[8] Philosopher and historian of economic thought Philip Mirowksi characterizes Flood as a visionary when it comes to the development of what he refers to as "cyborg initiatives"—that is, the melding of cybernetic developments (computational economics, game theory, software) to the sociopolitical sphere.

RAND conferences such as "The Theory of Organization" and its summer sessions on logistics (where many war games were developed) were part of the overall security society search for ways to incorporate communication and command into scientifically governed social hierarchies. Even though the search for such universally intrinsic hierarchies is the stuff of many academic theories, it was the military that first laid claim to an essential security *need* for such theories. The military would substantially develop this connection in and through the incorporation of scientists working in the fields of physics and biology.[9] At these conferences, Flood and a host of others would extend these connections to the formal sciences during and after the war. It was at the RAND organization theory conference of 1951 that Flood would articulate precisely the organizational work being invented—and radically imagined—to cement the civilian and military sectors in the generation of this new security initiative:

The objective of the theory of organization should be equally applicable to corporations, organisms, physical parts of the body or of animals, as well as to such things as family structure. Much stress [has been] placed on the need for versatility. Attempts at RAND have been made along "Robotology" lines. Robotology is not to be specifically animate or inanimate. Tests are being made of humans, of physical objects, of nerve networks, etc. Householder described the nerve network of the RAND TOAD, a mechanical contraption built at RAND that reacts in a predictable way to certain stimuli. Kleene put the nerve network of the Toad into mathematical form, and constructed existence proofs that other networks could similarly be put into mathematical form. Robot Sociology is being studied. Bales, at Harvard has constructed a set of postulates about leadership etc. Robots may be constructed that behave according to Bales' postulates. . . . Synthetic equivalents to human beings can be built and observed.[10]

According to Philip Mirowski, Flood's work in applied robotology con-stituted an immense advancement by breaking through the inherent limitations of work being done by social scientists "still mired in their precomputational preoccupations."[11] Mirowski refers to this develop-ment under the general term of *cyborg science*, which he takes from the work of Donna Haraway. Flood's articulation of the "need" to develop a "robotology"—a general philosophical system and usable material apparatus that would combine corporations, human and animal bod-ies, the structure of the family, and so on, all within a generalized and predictable "nerve network"—indicates Mirowski's apt appropriation of the term.

Mirowski's important historical analysis nonetheless limits these devel-opments to the Cold War period. Manuel De Landa reveals, however, that such developments were part of the actualization of large organizational structures that began at least a century earlier:

The first step in this migration of control from humans to machines was part of a long historical process that began with the first attempts at a rationalized division of labor. Although this process received its main momentum from the efforts of military engineers, it was also developed in certain civilian sectors, the tex-tile industry, for example. The earliest form of software was a set of pattern-weaving procedures stored in the form of holes punched in paper cards. This was the automated loom introduced by Jacquard in 1805. His device effectively withdrew control from the weaving process from the human workers and transferred it to the hard-ware of the machine. This was the beginning of a new migration.[12]

De Landa's analysis focalizes the desire to remove the human element from the chain of production. However, this emphasis overlooks the important role that this development plays in the larger ideological quest to extend human subjectivity into the realm of the geological. Centralizing the control of the production process by shortening the chain of command, coupling this with the general establishment of interchangeability in the modes of production (of firearms, vehicles, etc.), and, as Taylor suggested, streamlining (rationalizing) the labor process all have the goal of extending "man" in "man the machine." With Flood's robotology, these extensions of the human into realms of greater and greater proportions successfully transform "machinic man" into "man the machine of targeting and speed." The extensions of human influence that first began in the era of enclosures, which were further augmented during the Industrial Revolution, were now being extended even further with the help of mathematics and advanced computational electronic systems. It is this genealogy—of a specific militarized human development—that makes possible the constitution of the human as a geological force.

A little over a decade later—when such cybernetic systems had come into actual existence—Flood would extend the possibilities of neomilitary robotology, advocating its extension at the international level:

> Guided missiles, moon-crawlers, and various other cybernetic
> systems that now exist, provide us with relatively simple examples
> of inductive machines, and it is in this direction that I predict the
> next major breakthrough in the inevitable sequence of intelligent
> machines. . . . [T]he future of intelligent machines depends vitally
> upon progress in developing valid and practical theories of social
> relationships. Fortunately, intelligent machines are already help-
> ing us in the scientific work to understand social processes and to
> develop proper theories of economics, psychology, sociology and
> international relations.[13]

Flood's conception of a superstructural machine mathematically orchestrating the security and organization of human populations is an example of the human-as-geological-force imaginary that marks the status of the human in the anthropocene. Within the simple ideological parameters of scientific progress, the presence of this force and the potentially destructive nature of its performance are difficult to see. Flood's unidirectional pattern of thought, however, is of a piece with the "organic movement" of scientific discovery we saw in Frisch's and Oppenheimer's enchantment with nuclear energy: both are forms of intelligibility orchestrated by the hyperawareness, in an age of excessive information insecurity, of the need for greater apparatuses of infallible security. Flood's search reaches its summit with the creation of "intelligent machines," which he imagines as a kind of formal version of atom splitting that will generate a self-perpetuating system of inherently logical support systems in the human (war) effort to "understand social processes." These in turn further extend the scale of human intellectual pretensions to act as a geological force. From such efforts, no doubt, come the dream of a drone-protected society, of which the unmanned aerial vehicles armed with "hellfire" missiles are only the most recent conspicuous example.

Manuel De Landa's Genealogy of Military Environmentality

In this section, I look more closely at the events that made possible the general orchestration of the transformation of the human into a geological agent. As I indicated briefly in the introduction, the military has always been deeply integrated in environmental matters in terms of geopolitics, land relations, the development of nations, and contestations over land and sea in the colonial and postcolonial conquest of space. As the history of enclosure reveals, the privatizing of environments for purposes of security and national advancement requires extensive military commitment. In the twenty-first-century era of climate change, the enclosure of ecosystems for purposes of "energy security" now stands as the decisive "peacetime" battle waged by nation-states and global organizations across the planet. But at the same time, these efforts to secure environments reach their full realization, their effects extend beyond their intended targets, generating accidents on a global scale. Here I examine in detail the central constitutive role that militarized environmentality plays in the development and administration of an ecological worldview and in the constitution of a new image of the human that takes the place of what had formerly been identified as nonhuman and sublime. As I argued in chapter 2, the customary image of the human was fundamentally transformed when the enclosure movement and the new judicial system of common law went to war against the image of the human as an inhabitant. In turn, this movement formed a new image of the human as standing over and against a secured/privatized ecosystem, introducing a particular idea of environmental transcendence that began the historical process of extending the scale of human subjectivity into an imperial register. Such developments were made possible by particular military influences that guided technological development in the nineteenth century.

The enclosure movement established a new form of human subjectivity, one that put humanity in line with a more security-based orientation toward the environment. On a discursive level, the movement generated an extensive "structure of national feeling" (to invoke Raymond Williams's instructive phrase) about environments in the colonial domain. During the long contestatory historical development of this discourse, military metaphors played a key role in the development of a scientific attitude toward advanced land "improvements." As I've argued elsewhere, this is manifested in the most fundamental of geographic signifiers: the periodical "table," the foreign "territory," the enclosed "field," the designed "landscape." Each of these terms, as Foucault points out, reflects a new *administration* of knowledge that imbricates power to space.[14] However, this administrative dynamic reflects a military dispersion of power that makes this particular production of knowledge an essential part of the movement's targeting and securing of the new (enclosed) spatial relations defining the colonial landscape, an "information ballistics" of targeting and securing space that flows across multiple sites along the chain of being. "Territory" names a juridical-political dispersion of power: a mapped area controlled by a lord, a military commander, an imperial surveyor, a governor, or a nation-state. "Field" denotes an economic-military dispersion: the landlord at home (or the commander abroad) who encloses a space of land, turning it into his or her field in order to expand his or her individual (and national, in the case of the military commander) sovereignty and income (the stock of grain or sheep). "Landscape" indicates a military-aesthetic enframing: the comportment of land to an artist's image of organic beauty that also parallels the enframing of heterogeneous spaces by units of patrol so as to produce a "snapshot" that can be analyzed in the command center. It is during the eighteenth, nineteenth, and twentieth centuries when an entire constellation of terms such as these become generalized. They reveal the development of a *metaphorics* of a military unification that guides future scientific explorations into the procedures for securing territories—turning the heterogeneous singularities of land into comprehendible, quantifiable, and useable space.

It is important to remember that this metaphorics is not to be understood as a mere representational tool that is used to describe an organic or empirical state of affairs. Such terms, rather, can more genuinely be understood as a kind of technological apparatus through which a people of a national security state are constituted and imbricated in the security society's project of waging (peacetime) war on the environment. This metaphorics, in turn, is adopted as a usable and scientific terminology for determining in advance the possibilities of thinking about spatial relations. More than simply a mass of metaphors, they begin to *set the terms of interpretation*, to establish new comprehended "truths." These new truths are, in turn, administered on national and international levels, forming a generalized environmentality that naturalizes forms of development and makes possible the general transformation of Nature into Energy. These transformations lead me to the important work of Manuel De Landa.

De Landa is rarely invoked in the context of ecocriticism, in part because his work focuses mostly on establishing intersections between Deleuze, Baudrillard, science studies, and theories of complex systems and artificial life (though he occasionally touches on the relationship between agricultural development, trade, colonialism, and conquest).[15] In terms of his analysis of science and systems of security, I focus here mainly on his *War in the Age of Intelligent Machines*, in which he traces the contemporary history of scientific developments in technology and their essential relation to warfare. Combining the work of the Foucault of *Discipline and Punish* and Deleuze and Guattari's concept of the war machine, De Landa argues that the relationship between human beings, technology, and information production has undergone a profound shift in the modern era. Though part of the book provides a Braudelian, long-scale historical analysis, he concentrates the bulk of his critical focus on the nineteenth and twentieth centuries. The importance of this particular work for my purpose here lies in its attention to larger structural forces (the term he uses, after Deleuze, is "machinic") that have an impact on the development of human history. These structures are often human influenced or produced, but in the wake of years of naturalization, they begin to take on a life of their own. At such a point, they are interpreted into human epistemology and ontology as natural or self-evident. The enclosure movement is a prime example: widely resisted during its initial formations as an artificial and threatening form of existence, after years of development and expansion, and through its erasure of alternative forms of existence, it came to be taken as an empirical fact of sociopolitical reality. To clarify, the point in examining De Landa's theorization of structural or "machinic" forces is not to show that current human subjectivity must throw off its artificial, human-produced understanding of physical forces in order to return to an authentic understanding of subjectivity, as if it were "a product of nature." Rather, the point is to unconceal the ontological makeup of these structural forces (their essence of intelligibility) and their gradual development into a "natural" state of existence. In this sense, De Landa offers us a sense of "understanding" called for by Chakrabarty.

De Landa is also important because he places central importance on the military as a formative element in developments in modern technology and science—developments that in turn impact sociopolitical formations in the modern era. He traces a different genealogy almost entirely ignored in postcolonial historiographical critiques of modern scientific processes, which tend to focalize the Enlightenment lineage of scientific progression. Instead of Cartesian influences on subjectivity, taxonomic developments and social Darwinist effects on race relations, or Enlightenment developments in "Reason" impacting the colonial other (influences that are well acknowledged in postcolonial studies), De Landa traces the singular influence of military intelligibility on modern technological development. In this sense, his work presents a productive alternative to current periodizations of colonial and ecological historiographies.

War in the Age of Intelligent Machines begins with one of the most traditional historico-ecological periodizations: the shift from hunter-gatherer cultural formations to agricultural formations. De Landa uses the traditional terminology of world historians, who locate the dawn of "modern man" (the historical human proper) in the transformation brought about by the agricultural revolution. But he recasts this historical development in terms of war: "Early human hunters were part of the natural world and therefore were connected to this animal machinic phylum. Their early hunting tools and habits could have evolved continuously in a natural way from this portion of the phylum. But hunting tools did not become weapons of war as part of animal evolution. The war machine, where a tool becomes a weapon, involved social components, like the economic mechanisms of pastoral life, and these belong to human history proper" (38).

De Landa's marking of this fundamental change is grounded in a critical theorization of the potency of speed as a new and primary ontological force. (Speed, as we shall see, is a significant concern of the war machine.) The source of De Landa's discussion of speed is Arthur Iberall, the American physicist and engineer and a pioneer of "homeokinetics," the physics of complex, total systems. Iberall writes,

> The transition from hunter-gatherer to agrarian society, although profound, was not difficult for humans. To accomplish the change, they first had to adapt their behavior to that of the species they wished to domesticate (e.g., as nomads following migratory herds). It was then necessary for them to put selection pressures on the reproduction of the chosen species to accelerate their adaptation toward human requirements. Results were produced on a time scale much shorter than that of random, evolutionary natural selection. Human epigenetic [cultural] processes have a time scale of perhaps 1000 to 2000 years and are about 100 to 1000 times faster than genetic evolutionary processes at the species level.[16]

The first aspect to notice here is of course the shift from an ecosystemic system (humans as part of an environment and borne along by that environment) to an anthropocentric system (which adapts nature "toward human requirements"). More important, once this shift occurs, a new form of movement or "speed" comes into existence, one that can be identified as operating differently from the migratory demands of human–ecological movements that were in existence prior to this development. It is therefore necessary to make a distinction between different rates of movement: one rate of movement existing in an open-relational or ecological context and *speed*, which, against the former, would refer to a human-centered orchestration of movement, an "acceleration," to use Iberall's term, that is trans- or counterecosystemic in nature.

The second aspect concerns the idea of "epigenetic." Iberall uses the term to differentiate between human processes and ecological or "natural" processes. The discipline of epigenomics in evolutionary biology somewhat matches this particular use of the term. For instance, studies have argued that changes in environmental factors have a greater impact in small, short-lived species than in longer-living species, like humans. In this sense, humans accustomed to a certain counterecological speed are not in the position of needing to open themselves to new relations as they encounter new environments. The demand for a certain "target velocity"

requires that their environments be adapted to their sustained level of biological standing, which grows more homogenized and steady across time
and space.[17]

Consider the following application of the concept of epigenetic memory in plant biology. Genes in plants are affected in the long term by
environmental factors. Yearly changes in temperature, for instance, are
"coded" in the genes of certain plants, even though there is no change
to the genomic sequence. This is how plants "remember" when winter
is going to come to an end and thus appropriately time their pollina-
tion and seed dispersal.[18] The concept of epigenetic memory thus offers
an understanding of a precise and singular relation to an ecosystem and
its temporal coordinates. We might, consequently, productively identify
the "anthropocentric human" (the human acting as a geological force)
as the species that transcends or goes beyond the epigenetic limit of its
specific environments—destroying the distinct, heterogeneous temporality of ecosystems in favor of a war-oriented "genetic disposition." This
war-oriented "genetics," it should be emphasized, is only one among many
possible dispositions. It owes its prevalence, as we've seen, to specific historically and politically influenced patterns of development. In turn, this
image of the human also speeds up particular processes by developing a
pattern that, when normalized, transcends the distinct qualities that other
organisms attempt to maintain in their epigenetic memories: the militarized human moves from heterogeneous movements to a unidirectional
and accelerating rate of speed.

The Higher-Order Motor: The Military Holy Grail

Speed, for obvious reasons, is an important factor for the war machine.
Its successful incorporation into the chain of command leads De Landa
to consider the military's search for what one might call its "Holy Grail":
a behavioral disposition and a form of speed that can be universalized
across time and space. According to De Landa, from the eighteenth to
the twentieth centuries, Western military organizations underwent a
fundamental transformation in their structure and development. Part of
this development has been analyzed in the classic work of Michel Foucault, specifically the rise of "disciplinary societies" and the constitution
of the "docile body" as the primary form of these societies' form of individuation. Foucault, however, does not fully focalize the military site of

production and its relation to science, nor does he discuss the military outside of a (mostly) generalized French context (though he does mark its ontological and rhetorical importance). Like the enclosure movement, European developments in military systemization influenced American security methodologies. French and German military systems had a significant effect, for instance, on American military and on Western military production in general. Beginning in the mid-eighteenth century, innovations in standardization swept across the Western war machine: from the normalization of guns and gear in the French artillery by Jean Baptiste Vacquette de Gribeauval, to the adoption by the U.S. Army in the mid-1790s of Honoré Blanc's identical parts in small arms, to the general military uniformity movement promulgated by Eli Whitney, to the more expansive military-technological innovations during the Industrial Revolution, and finally to the Fordist groundswell in mass production.

These and other connections reveal the extent to which organized uniformity came to be a central and growing mechanical concern for military organizations in the nineteenth century.[19] Such developments in standardization cannot be explained away as simply evolutionary advancements in the history of modern warfare. They reflect the development of a new historical ground plan and an attempt to universalize a form of existence that will demand increasing measures of security. Within the register specific to the relation of military production and technoscientific developments from the eighteenth century to the present, this access manifests itself in terms of processes and movements that are understood to arise out of Nature itself.

One of the coveted treasures that motivates technomilitary development, argues De Landa, is the concept of the universal "higher order": the name for a hidden ontological presence or powerful force lurking behind human and nonhuman activity. He finds that the fantasy of a "higher order" not only informs the search for uniformity but also generates a specific constitution of the soldier and the officer. The influence of this idea is extensive and continues to heavily govern military theory—so much so, argues De Landa, that it becomes the driving force behind the search for artificial intelligence in the late twentieth century. De Landa writes, the "history of military applications" must be thought of in terms of "the effects of technology on the military, understood here as being itself a coherent 'higher level' machine: a machine that actually integrates men, tools and weapons as if they were no more than the machine's components."[20]

The ability of a self-organizing event—a singularity that changes the course in which environments (human, mechanical, ecological) rise and develop—to take on higher-level or higher-order activations *names the moment at which a certain mode of production formerly limited to a specific site or field of creativity expands to enchain multiple sites*, thus becoming a globalized and seemingly organic mode of production. It is the moment when war takes on the pretension of being the basis of existence. In such a way, military technology sets itself up as a field of knowledge production concerned with the fundamental questions of the universe—fundamental questions that can lead the scientist and the military leader on the path to find the ultimate, natural foundations for universal security itself (seen in the fantasy of an innate order of existence in the splitting of the atom and in Sagarin's fantasy of "natural security").

De Landa first traces out the development of a higher-order machine of war at the site of cryptology. This development begins with the creation of cipher machines—the tool that eventually develops into the modern personal computer: "The first modern computers were assembled in the crucible of World War II, in the heat of several arms races: the cryptological race against the cipher machines of Nazi Germany and Japan and the race against German scientists to build the first atomic bomb. The war produced not only new machines, but also forged new bonds between the scientific and military communities" (5). Here we can begin to understand (and not simply "explain") the essential relation between the war machine and acceleration. Militarized ontology has its essence in the enforcement of a "target velocity," designed to replace heterogeneous forms of motion and movement and teleologically direct them toward the generation of a process that "cannot be stopped." This empirical need to be the first to solve the problem of encrypted messages, however, did not remain distinctive to the problem of breaking enemy codes. The intense pressure on the allied forces to crack codes generated a magnitude of force that angled an explanatory form of science (which it concurrently developed) to military concerns: "Never before had science been applied at so grand a scale to such a variety of warfare problems" (5). That particular form of science and the objects it produced formed a kind of technological "swell" that impacted both political and social production on a massive scale. The result of this collaboration, the discipline known as *operations research*, has evolved in the hands of cold warriors and think tanks into the more inclusive *management science* (systems analysis), which in effect transfers the command and control structures of military logistics to "the rest of society and the economy." At such a point, human agency and the larger-than-human structures become *effects* of the "higher-order" motor connected to the larger groundswell generated through the intense pressure of science and the military working in conjunction to solve the problem of mathematical encryption. De Landa reveals how this new organizational process arises out of the pressures of a historical occasion that brings about a unique machine, a new set of procedures that expand in multiple directions: from the immediate context of decoding encrypted material to an entire field of logistics that begin to take up residence in the economy and social system at large. The rise of this "higher-order motor" is thus both a new singularity and an event derivative of preexisting ideological structures. This "singularity" that generates a new assemblage—information technology—should, despite its unique character, equally be seen as part and parcel of a much longer history of being (*Seinsgeschichte*) of metaphysics—of a mode of thinking and representation oriented toward a particular end: totalized comprehension and mastery (total system theories).

The Higher-Order Motor and the "Projectile Wall": From Nomadism to Sedentariness

The prevalence of the military's search for a higher-order machine should be given the greatest possible attention. It acquires such central importance in the security society that its complex historical influences become reified into the form of a historyless fetish, into a globalized, panoptic *idea*, polyvalent in its manifestations and uses for neoliberal gain. If we were to assign an ontological "core" to this idea, it would be the unidirectional dynamic of targeting. But it should be remembered that this targeting action is not an empirically verifiable phenomenon: its ontological justification can only be "grounded" in the fabrication of an insecurity—an enemy other it must constantly generate, seek out, target, and eliminate.

De Landa reveals the original centrality of this targeting dynamic when he focalizes the shift from nomadic forms of conflict to the first developments in sedentary (state-organized) forms of war—what Virilio would come to identify in the 1970s as "total war": "Despite the many improvements that the Romans made to the phalanx concept, the nomad paradigm remained the most successful way of waging war until the late fifteenth century."[21] The phenomenon that made sedentary forms of war possible was the state institutionalization of gunpowder as a structural force. He interprets this emergence of gunpowder within the state apparatus as a symptom of a new military desire—a higher-order motor—which manifests itself in the successful "corning" (refining) of gunpowder several centuries after its initial introduction into the European military community. In other words, the being of gunpowder at its first historical appearance is not the same as what it ultimately becomes. Gunpowder must first take on an acceleration—a shift to an anthropocentric, war-oriented speed—as it becomes inserted into an antinomadic system of development and production not in place at the time of its initial introduction: "At that point [after centuries of 'corning'], the appearance of a new breed of machines—gunpowder-based mobile artillery—decided the battle against the warriors of the Steppes. The sedentary way of war would now begin to dominate the martial landscape" (12). Gunpowder became a new device, different from its initial material conception—different in its very ontological essence—and that new device helped make possible the establishment of a sedentary and more security-oriented mode of existence. In turn, the ability of the extensive sedentary state to devote large resources of economic investment into the production of static cannonry (which made possible an exclusive form of speed that could be accelerated) began to overpower the heterogeneous mobility or "raw speed" of nomadic cavalry: "Walls of metallic projectiles produced by volley fire triumphed over raw speed and surprise" (12).

Here De Landa and Paul Virilio come together in their theoretical analysis of the impact of war on contemporary existence. Both De Landa and Virilio locate the origin of this new ground plan of standardization, and the logistics that serve as the highest level of its cultivation, in precisely this development in artillery, which introduces a single, state-administered form of speed. Tellingly, and important for my argument here, both theorists notice that from this military artillery program arises a new phenomenon: the "projectile wall." The projectile wall is defined as the artifice that makes possible the amplification and spreading of militarized targeting speed across a great surface area—an early form of "shock and awe." De Landa locates this development in the "American system of manufacturing" (32). (Virilio, as we will look at shortly, ties it to scientific advances in optics—specifically the American and European film industry of the early twentieth century.) The idea of the secure/insecure opposition informing sedentary existence and its successful manufacturing of a "projectile wall" designed to conquer nomadic existence is not considered a colonizing force. Rather, it is troped as one of the hidden truths of existence, as a natural force stemming from an axiomatic universal order—a providential "higher-order motor" that signifies civilizational progress. Civilizational progress in this sense is a discovery made possible by coupling the representational apparatuses of human-historical progression to the targeting apparatus of the war machine. Unlike the nomad, the state has the economic power to *arsenalize* the potential of gunpowder's explosive force, introducing a new trajectory that begins to dominate historical processes. Gunpowder as an entity begins to extract itself from an evolutionary progression that can be identified with its gradual migration from China and erratic development in Europe. It becomes something else: a singular force or motor that pushes an entirely new set of formations into existence. It is at the same time an unbridled energy and a highly submissive energy. No longer a set of chemical reactions associated with a discrete phenomenon and more than an advantage or new potentiality within the known game of war, "gunpowder" becomes synonymous with a new ground of existence, a new generalized order setting in place a whole innovative set of inventions dependent on one another. It turns the nomadic relation to land and movement—land as the name for that which supports an unenclosed inhabitancy—into *a technical problem needing to be solved* by means of subjugation and centralization. The entire meaning of the ecosystem changes at this point, brought within a logistics governed by forces of a single, militarized speed and coconstituent of a targeting system. The environment as a support for inhabitancy disappears, and what appears in the place of this is an entity essentially recalcitrant and in need of regulatory mechanisms that will unlock its furtive capacity for development. Here we can begin to see the full irony of an invention originally valued for its role in the production of fireworks and celebration. Gunpowder now becomes the sign of an aspect of Nature honed to its full capacity, a force suddenly "revealing" its true potential. As a material element, gunpowder has nothing to do with supporting existence and interactive habitations. Instead it allows one to colonize habitations, recoding them as mapped geographies of an unending projection of enemy-others.

The shift to a sedentary stance transforms warfare into a matter of mastering an entire spatial region in a single instant of time. If gunpowder marks a limit that changes the structure of warfare and civilization, as De Landa suggests, then it also generalizes the production of warfare, in and through the new paradigm of visualizing the entire ecosystem as a spatial grid that unfolds the landscape as a targeting surface—what amounts to, in other words, a logistical and ballistic-oriented system of enclosure that floods across whole regions newly understood from the basis of needing to be secured and controlled.

Nature, at such a point, ceases to have anything to do with sustainability; it is hardwired into a targeting apparatus that understands advancement as the increasing capacity to be hyperaware. Even if nature were coded anthropocentrically before this fundamental shift in waging war simply as a raw resource for human consumption in terms of food and shelter, it would not necessarily bring an ecosystem into a discourse of enclosure and a capitalist mode of production, as we know from the long centuries of sustainable human–land relations. But when mass amounts of firepower are demanded at the level of the state, the environment takes on a different, secondary status in relation to the generalized economy of warfare and the resources it demands. Land becomes a matter of *logistics*: "A given war machine is composed of a hierarchy of levels, a series of components operating at successively higher levels of physical scale and organization. At the lowest level, there are weapons. . . . One level above we have tactics. . . . The next level is one of strategy. . . . Finally, we reach the level of logistics, the art of military procurement and supply, which may be seen as the assembling of war and the resources of the planet (fodder, grain, industrial might) that make it possible" (23). This development of a hierarchy goes hand in hand with the shift to sedentariness. It reduces nature (minerals and grain) to an object that only obtains meaning as the "raw materiality" needed for energizing the "next level up the scale." The "top" level of this scale—logistics—culminates in the establishment of the post–World War II military industrial complex, which, in its need to reorganize life around the transformation of knowledge into "information," generates the new ground plan of the information age and thus the globalization of thought that reconstellates the environment as an entity in need of mastery. In turn, the accidents that inevitably follow in the wake of any targeting exercise multiply to eventually become the principle substance of existence itself. Climate change, resituated within this countergenealogy, reveals itself to be the example of the accident transformed into the primary existential character of a historical age.

We can extend De Landa's theorization of the fifteenth-century projectile wall to more modern historical examples. One of the first acts leading to the projectile wall's institutionalization occurred in the eighteenth century with the transformation of the nomadic Jäger militiamen and the German Jäger rifle the militiamen used in combat. The Jäger were a type of lower-middle-class soldier recruited by Frederick the Great during the Seven Years War. They were skilled riflemen and famous for their success on the battlefield, but their weapons and their movements were nonstandardized. Jäger rifles were heavily personalized and varied in style, yet the singular power of this weapon on the battlefield caught the attention of military leaders in England and the United States. Toward the end of the eighteenth century, the Jäger rifle underwent a uniform shortening. This made the rifles easier to load, which was more practical in skirmish combat. At the same time, Jäger rifles were exported to a number of countries, including England, which then transported them to the States during the U.S. Revolutionary War. In 1810, the Jägerbüchse (hunter's rifle) became one of the first weapons to be standardized after the defeat of the Prussian army at Jena-Auerstadt.[22] Eventually, the Jäger rifle was displaced by even more uniform ballistics weapons, but its simultaneous homogenization and insertion into several state military apparatuses is a symptom of one of the first leaps to a standardized "wall" of weapons. This simultaneous production of uniformity and universalization would come to be the norm in the new Enlightenment-age military machine. It is thus much more than a "drive to uniformity," as De Landa defines the process (31). Management becomes "scientific," and that science is governed by an endeavor to accommodate difference into a dependable, useful, politically conservative yet highly productive system of operations. The Jäger rifle is suddenly understood to be more primitive in its original formation. Uniformity becomes the answer to unleashing the true potential, a potential that, when fully powered, enters the domain of the universal. This, in turn, generates the ultimate prize of a "higher-order motor" all the more efficient for its self-contained and self-generating capacity.

De Landa historically locates the decisive military colonization of the sociopolitical system in the nineteenth century, when, he argues, the singularity of mechanized systems took over multiple sites of social production.[23] (As I argued in chapter 2, the enclosure movement prepared the way for this development more than a century before.) This transformation first occurred with the introduction of military mass production—a material fact that was the result not of industrialism but of the war machine. It was in the American and French army system where the "standardization and routinization of manufacturing practices were first introduced."[24] Standardization arose as a necessary goal to solve the "new problem" of how to create weapons whose parts can be interchangeable. The fact that interchangeability arose as a "problem" at all is symptomatic of the shift from a nomadic inhabitancy to a sedentary form of existence. At such a point, the manufacturing process must also generate "deskilled" workers (Deleuze's term, or what Foucault called "docile bodies") that match the "standing reserve" of the preformed, interchangeable parts of the weapons being constructed. In this sense, labor is inserted into an economy of "rationalization" originally demanded by a military protocol and not, as is commonly thought, because of the Ford/Taylor lineage (Taylor, for instance, developed his ideas for the assembly line by observing military camps and their activities).

Evidence suggests that this move to a more highly controlled, sedentary form of waging war arose in tandem with an ontological shift to security as the basis of existence. De Landa reveals, for instance, that the expansive push for uniformity originated in the Ordnance Department (whose essential purpose was management—the supervision of procedures so as to achieve an optimum of efficiency).[25] Any successful, long-term uniformity program would need to be orchestrated by much more than a rudimentary level of monitoring practices, and it was precisely a new military practice of monitoring that eventually became one of modernity's bases of existence, an administrative ordering "later transmitted to the civilian industry via the contract system."[26] The overriding problem that a standardizing war machine faces in addition to security is the development of structures that can productively control the flow of social spaces across great geographical distances. Solving this problem immediately required a lessening in the gap between the military and the civilian sectors. The distinct and heterogeneous movements that defined the original Jäger corps had to be replaced by an overall management of flows that could bring heterogeneous sites of production and human movement under a centralized and uniform flow of command that grew increasingly rigid: "military orders became 'frozen in steel'" (32). The very material reality of distinct and differential relations between humans and the material world had to be transformed: "The military sought to avoid dependence on human skills and launched a scientific investigation not of the singular properties of metals, but rather their uniform properties. The generating of a given shape by following the local accidents of a given piece of material was replaced by schemes for insuring the imposition of a 'uniform shape' across all lines of variation" (32). This "dissatisfaction" or "disinterest" in human skills in favor of "uniform properties" in the physical world *is one of the first signs of the military's interest in locating a "geological" order in nature that would exceed and replace human limitations.* The demand for uniformity brings into existence a militarized human "collective," one that transcends or overcodes the heterogeneity of nomadic movement, transforming it into a globalized, geological force. This new "order," thought to be a release of an optimum force in both nature and humans, extends the naturalized order of gunpowder that precedes it, transforming humans into a higher-order "force of nature."

Numerical Control: Environmentality in the Formal Sciences

The targeting of the uniform flows of materiality obviously had to be coupled with a similar style of uniform "compliance" on the part of the human body—that is, the military personnel who would put to work the homogenous weapons being formed. As Foucault has shown, this organized uniformity was the pride of Napoleon's army, and the "docile body" was the form of human subjectivity refashioned into the disciplined machine of a centrally organized movement. Foucault productively connects this polyvalent dissemination of the docile body to sites beyond the army: schools, hospitals, factories, and other social spaces. De Landa takes us one step further in this expanding network and makes the even more productive connection (in terms of my focus here) between the organized factory system of Taylorism and the creation of the twentieth-century "computer age," which has its origins in the need to monitor "information" and "numbers": "The modern representative of the last century's drive toward uniformity is the system of Numerical Control (NC), the product of research funded by the Air Force in the 1950s. Fueled by the Korean War, the NC allowed the translation of parts specifications into *mathematical information*" (33). Here the new network of uniformity takes on an empirical grounding with the addition of mathematics, the system defined by its dealings with what is "known in advance" and "already known," as the original Greek *ta mathemata* connotes.[27] In other words, mathematics becomes a viable solution for introducing an infallible level of certainty

within the security problematic: as a respected formal science, it constitutes, in its particular mode of intelligibility, a trustworthy ground of
knowledge production and a form of *explaining* that requires no Gadamerian understanding.

The introduction of math into the network of militarized uniformity
may appear initially as a late or more "modern" development. But it owes
nothing to the idea of a progression natural to technological development.
Math becomes important because of its affiliation to this assemblage. The
essential aspect of this field lies in its ability to establish, on a representational level (numbers representing the motion of units of mass and their
spatiotemporal coordinates), the tightest possible connection to the largest expanse of materiality (the physical universe) coupled with the ability
to map the maximum quantity of detail. Hence the rise and importance
of "numerical control" as a military concern and the various fields and
subfields of statistical information gathering—the development of a field
of inquiry based on the representation in advance of what is already
(empirically—that is, ontically) known. Statistics (the control of empirical reality through the use of numbers as a representational language—as
a form of *vorstellung*, to invoke Heidegger's important term: a form of
representation that acts as a "setting-up-before," "setting-in-place," and
"pinning-down")[28] easily meshes with the war machine because it shares
an affinity with military desire (representing so as to identify and target).
The military historian David Noble reveals how this numerical reduction
of diversity in the statistical "command chain" erased a certain "human
intervention": "The vision of the architects of the NC [numerical control]
revolution involved much more than the automatic machining of complex parts; it meant the elimination of human intervention—a shortening
of the chain of command—and the reduction of the remaining people to
unskilled, routine, and closely regulated tasks."[29] With the rise of "computing," the "central command" of the work force—management—could
directly "bypass the worker and communicate directly to the machine via
tapes or direct computer links. The machine itself can thereafter pace and
discipline the worker."[30] The "computer" was thus pushed into existence
in order not only to calculate but also, and more important, to target and
to do so by transforming human agency into a more-than-human geological force.

These developments brought about an essential transformation
of the human from an inhabitant (and the custom system of law) into a

subject marked by a process of *identification-individualization* (enclosed-standardized). Enfolded in advance within the security discourse demand-for-knowing, humans and the ecosystem are henceforth stipulated in advance so as to appear *primarily* in this identification mode—a form of identification that does not necessarily nor fundamentally counter the rise of identity politics. Nor does this mode of identification necessarily reveal its essential militarized logic in the poststructuralist deconstruction of the Cartesian, sovereign subject. It is not a "decentering" of a subject that opens a radically new form of freedom from the oppression of metaphysics but instead an understanding and critique of the establishment of subjects produced in and through the development of statistical-targeting assemblages: anticipation, not sovereign meditation, stands at the core of the technologico-military capturing of the subject. The understanding of a self as an entity having an internal apparatus for meditation is a vague and incomplete manner to understand how capturing mechanisms operate. The deconstruction of the sovereign human subject no doubt offers the advantage of demystifying ideological practices that attempt to mass-produce the idea of a subject beyond all ideology. But if the essential character of the anticipatory character of the age is overlooked, the critique of the sovereign subject remains incomplete, for such a critique still leaves intact the notion of an exteriorized subject that establishes, and is part of, an anticipated and statistically mapped exterior world.

The subject of the modern, militarized era does not statistically map and target existence as if this mapping were an option among many that one simply chose; the modern subject arises out of and is borne along by statistics and targeting. As De Landa shows, the military has sought in the past to maintain control over the institutionalized structures of uniformity, transforming information production and accumulation into what we might call a "statistical shield." This demand, as I've been attempting to show, is polyvalent and occurs in times of peace as well as in times of war:

> The system of NC is just one element of the Air Force's dream of the totally computer-controlled factory. . . . The birth of microcomputers should, in theory, allow workers to regain some control over the process of allowing them to program and edit the machines themselves. But these and other technological possibilities are being blocked by the military which sees in alternative man-machine interfaces a threat to its tightening grip. As the last

two great wars have shown, victory goes to the nation most capable of mobilizing its industrial might. Wars have come to depend more on huge logistic orchestration of effort than on tactical or strategic innovations. Imposing a tight control and command grid on peacetime production is seen to be the best way to prepare for wartime resource mobilization. Creative interaction with computers, although capable of increasing productivity, is seen as a threat to the perpetual state of readiness for combat that has characterized the Cold War years.[31]

As the sedentary discourse of statistical analysis develops in new directions, expanding its shield in new material forms of production, it releases singular forms of energy and singular directional flows of that energy that escape its defining parameters. The precious object of the "higher-order motor" of control begins to escape the military's grasp. The need to regulate and monitor the production process becomes more expansive and complex, and the need to establish tighter connections between all forms generative activity and the logistical "statistics shield" becomes more of an urgent matter of a potential insecurity needing to be securitized, in turn generating a desire for an absolute form of speed capable of anticipating potential threats.

Absolute Speed and Relative Speed: The Pacing of a Horse

De Landa offers a salient example of the danger inherent to the erasure of multiple singularities by marking the difference between "absolute speed" and "relative speed." In the field of fluid mechanics, relative speed is the speed of a moving body in relation to the viscosity of the medium in which the body flows, and it is measured in Reynolds numbers. The concept of the Reynolds number, as we will see, is extremely important, for, as De Landa argues, it pertains to "the biological machinery of the planet." (I develop this in more detail in the conclusion in terms of Jakob von Uexküll's *Umwelt*.) At a certain point, the ability of an object to support its movement in an ecosystem will become chaotic and unstable if it attempts to perform the same type of movement at a speed that introduces too much friction. De Landa uses the example of the shifting gaits of a horse. At a certain speed, a horse must shift its internal assemblage in order to increase its speed. It must shift from walking (a particular, contextual singular flow), to trotting (a different singular flow), to cantering, and finally to galloping. Each of these are motions but are distinctly different. If forced to trot at too fast or too slow a speed, the assemblage begins to become chaotic and unstable, eventually breaking down. De Landa articulates these singular, "self-organizing movements" in the following passage:

> Important changes take place in the nature of *animal* locomotion at different speeds. In terrestrial locomotion, changes in gait, from walking to trotting to running, occur at a critical point of speed for each species. . . . These critical points are not thresholds of absolute speed but a special kind of "relative" speed, the speed of a moving body relative to the viscosity of the medium, measured in Reynolds numbers.
>
> A Reynolds number is a simple ratio of two forces: the inertial forces of the moving body and the viscous forces of the medium through which it travels. . . . Reynolds numbers . . . have a very intimate connection with the machinic phylum. They can be used to identify the singularities in self-organizing processes. . . . More generally these thresholds of "relative speeds" divide the world into regions, across scales and flowing media. In these regions only certain kinds of animal locomotion machinery can develop, and the evolutionary success of a given design in one region does not imply its fitness in a separate region. . . . [A]ll the biological machinery of the planet has evolved following these thresholds.[32]

The double ecosystem is composed of the internal assemblage of the horse's corporeality and the external context of its specific location (a canter on land and through air may have to become a trot when moving through a flowing river). Likewise, each self-organizing process carries a relativity to its particular region, its ultimate exteriority, which delineates the potentials, contours, and limits of its movement and development.

The monomaniacal desire for an ideal speed (equivalent to a horse that would never need to shift gaits, no matter what its velocity) names a desire to rupture all barriers—in space and time. Its goal is the construction of a solitary contextuality, one that erases the wastefulness of duration that defines the "viscosity" of distance. This erasure of distance is essential to the production of ideal speed (the prized technologies of numerical control and military developments in instant telecommunication—on the battlefield, in the commercial sector, and in the mind of the citizen-soldier). Current technometaphysical politics replaces multiple, heterogeneous thresholds of speed with an absolute, universal speed because it is demanded by the very nature of closed-loop systems. It is a speed that has as its goal the rupture of all barriers, of all other forms of speed/movement. The difference between the gait of the trot and the gait of the canter, however, is *absolutely heterogeneous*: the movement of the canter will forever be distant from the movement of the trot. This is not a flaw to be "overcome" but the factor that gives them their distinctiveness. All these ideas and singular materialities are anathema to the war machine.

These developments in the areas of singularity, epigenetic memory, pacing, and an open-relational dynamic and, conversely, their minimization by transhistorical disposition, enclosure, speed, and the projectile wall not only operate at the levels of social and cultural constitution but also impact various levels of human consciousness. We can begin to see this impact in the context of De Landa's discussion of the indissoluble connections between early twentieth-century military investments in ballistics analysis and the rise of the computer industry. The computer as a form of technology ascended out of a certain mode of intelligibility that combined targeting (ballistics), information analysis (the calculating character of *vorstellung*), and human subjectivity. By enchaining multiple human subjects and working those subjects as a single machine, the first "computer" was in essence an expansion of the human into a larger, more potent force—that is, into the geological force that Chakrabarty identifies as the enhanced, climate-changing "nonhuman-human": "the term 'computer' meant a human operating [a] calculator. Organizing large groups of mostly female 'computers' for the performance of large-scale calculations was a task often performed in ballistic analysis and other military calculation-intensive operations"[33] (see Figures 14 and 15).

According to the ballistics analysis course required for women by the U.S. Army and given at the University of Pennsylvania, "Women often worked double or triple shifts in order to complete the ballistics calculations, which were published in booklets distributed to soldiers. They could be so precise that they calculated whether enemy soldiers were standing or laying in trenches."[34] Following the compulsion to form an apparatus that could successfully integrate degrees of information that lay beyond the capacity of a single soldier or command officer, the military amplified and

Figure 14. During World War II, women worked in tandem in order to become the first "computers" capable of handling large amounts of information necessary to crack enemy codes for the U.S. Ballistic Department. U.S. Army photo.

honed the data-accumulation potential of a group of humans to generate the first metahuman, higher-order ballistics machine—a human projectile wall of "intelligence" as the new social logic. This "intelligence" marked an insertion of consciousness into a system of exchange governed entirely by ballistic processes. The entire orientation of this system of knowledge production was geared toward finding the target as quickly and as efficiently as possible. The science of ballistics operates by reifying the complexities of an environment into "data" to be accumulated in a general system of logistical analysis. Out of this logistics arose the need for devices that would be able to compute at a much higher rate of performance than any human being. (The SAGE [Semi-Automatic Ground Environment] system was the first nationalized technological form of such a device.) This mode of consciousness exchange came to be prized over all other possible uses of intelligence. It constitutes a form of nonhuman human, geological consciousness that is trained to function at the highest speed possible, and,

Figure 15. Women working in the U.S. Ballistic Department, processing mass amounts of information with the first computers. Photograph by Harris and Ewing, Library of Congress Prints and Photographs Division [LC-DIG-hec-20580].

through honing these skills, they become "automatic" methods of explaining, easily "programmable" and "updatable" as new forms of technology emerge. These nineteenth- and twentieth-century militarized progressions began to reach their peak during World War II, under the general supervision of Vannevar Bush.

Vannevar Bush, National Defense, and Ballistics Research: Hardwiring the Accidental

The American engineer and preeminent science administrator Vannevar Bush was at the forefront of orchestrating new forms of a generalized war-machine intelligibility in the United States. An electrical engineer by training, Bush was the primary organizer of the Manhattan Project and oversaw the development of the first nuclear weapons. He was director of the heavily influential Office of Scientific Research and Development— the office created by Roosevelt's executive order, which took over the work of the National Defense Research Committee on June 28, 1941. He reported only to Roosevelt, was given practically unlimited access to funding and resources, and acted akin to a military leader in the science and technology community (*Time* magazine referred to him as the "General of Physics" in 1944). Bush had been concerned about the lack of cooperation between civilian scientists and the government during World War I. He almost single-handedly generated the links between academic researchers, the military, and the industry that would become the military industrial complex. Before America entered World War II, he had already been planning extensively to create what eventually became the National Defense Research Committee (NDRC). After Roosevelt agreed to its formation and gave him practically unlimited funding and political support, Bush appointed four leading scientists to the NDRC: James Conant (president of Harvard University), Karl Compton (president of MIT), Frank Jewett (president of the National Academy of Sciences), and Richard Tolman (dean of the graduate school at Caltech). Each of them was assigned to oversee specific military-scientific projects: Compton oversaw radar; Conant had chemistry and explosives; Jewett had armor and ordnance; and Tolman had patents and inventions. Like the NDRC, the primary purpose of the Office of Scientific Research and Development was the coordination and funneling of scientific research for military purposes. From 1941 to 1947, Bush channeled the research endeavors of some six thousand leading American scientists toward military development.[35] During and after World War II, he brought together members of the scientific community from a number of academic disciplines to work with the military to develop various forms of war "tools"—radar, missiles, and the atomic bomb—to advance the field of ballistics.

One of Bush's own ideas was to develop an early version of what would eventually become the World Wide Web. Called "memex" (a combination of "memory" and "index"), it was envisioned as a device that could mechanically combine the benefits of information access with "exceeding speed and flexibility." Memex was essentially a protohypertext meant to supplement and extend the limitations of human memory.[36] The idea for the machine arose from Bush's concern that scientific enterprises were oriented more toward destruction rather than understanding and that part of the problem lay in the lack of developing a collective, open-access memory. Bush positioned the sciences as the form of human production best suited to solve this problem. He thus failed to see the complicity of the untheorized ontology of technoscientific production with the ontology of the war machine. On an ontological level, both Bush and the military shared a fundamental desire to manipulate information at greater speeds and to funnel data into a higher-order motor or machine. The purpose of the memex machine, for instance, was to unite disparate "lower level" technologies under the control and rubric of a "higher level" of organized knowledge (to surpass processes of human memory recall). In order for both systems to generate a higher-order motor, knowledge and the patterns of thought dealing with that knowledge needed to remain at the level of *logic*—which emphasizes explanation at the expense of understanding. Bush's dream of the memex machine, therefore, follows in the footsteps of a certain form of pragmatism peculiar to the American nation-state. It is a form of apolitical, explanatory thinking oblivious to the rigidly orchestrated direction of its own thought.

This holy grail of a higher-order motor capable of dealing with information at great speeds would ultimately manifest itself in the early form of computational devices—initially in the form of human-technological machines and eventually in the nonhuman forms of computer technology we are so accustomed to using today. Because the war effort during World War II demanded that the United States and its allied forces beat Germany in the race to make the atomic bomb and to decipher complex mathematical cyphers, it was problematically (in Althusser's sense of the term) necessary for the computer to be developed beyond its initial projectile wall of women data processors. Both military offense and defense and the discourse and institution of security would have been impossible without the realization of this higher-order motor. In the worldwide war of capitalism versus communism, the United States began to develop various forms of security-driven assemblages that reflected the actual technological ability to fabricate a physical and totalizing Accidental apparatus, an ability that had previously eluded science. The Air Force SAGE program (chapter 1), an electronic structure, as we saw, designed to function as a "continental fortress" for the United States in the face of a potential Soviet threat, constituted the most ambitious attempt at that time to manufacture a fully operational form of the nonhuman human that would extend the ballistic imperative across an entire nation. Such a device materialized the image of "national security," which, for all technological intents and purposes, had existed only as a partially realized dream prior to its invention in 1958. As a project, it would bring together all the areas of the military industrial complex, all the various fields of scientific inquiry and military production that had been growing since the nineteenth century. In order for its "automatic environment" to become a reality, the national security system would have to become firmly embedded at the university level and subsequently buttressed by corporate contractors: IBM (which developed the computer hardware), the Burroughs corporation (which developed communications), Western Electric (which designed and constructed the program's twenty-three "Direction Center" buildings), and Lincoln Labs (which became the MITRE Corporation in 1958 and provided the necessary system integration).

More than the creation of an advanced system of total surveillance, the SAGE system made manifest (and cemented in technological materiality) a certain dream of combining and managing the technological administration of observation, targeting, specd, and power. With the SAGE system, the projectile wall of the nineteenth century was beginning to be brought to its inherent logical conclusion by slowly engendering its own "accident": the need to be entirely locked into a defensive "alert status"—an anxiety apparatus successfully functioning twenty-four hours a day and complete with "backup systems" (each facility had its second, backup SAGE computer, capable of being "fired up" at a moment's notice). If the coveted goal of the military was the reproduction of standardized weaponry a century earlier, then the goal of the post–World War II period was the construction of a workable and instantaneous system of "general alert"—a form of obsessive, pure interiority designed to constantly monitor the radicalized violence of an exterior it coconstituted as the monstrous fantasy that justified its existence.

The success of standardized and advanced systems of targeting, coupled with the destructive firepower of atomic and then nuclear weapons, was bringing an imperfectly formed concern into a new adulthood: security studies. "Security" was beginning to be hardwired into the performative identity of the neomilitary United States. This governing concern would soon expand from the military to the civilian sector, reaching an apex of heightened defensiveness in the post-9/11, George W. Bush era of *homo sacer* and the "state of exception." To put this in ecological terms, the idea of "sustainability" that directly speaks of a future was already in the stages of being overridden by the security demands of the present during the early Cold War period. The pragmatic and expedient solution of "adaptation" was being hardwired into the technological ground from which future experiments would owe their existence and potentials. The mode of existence governed by adaptation, therefore, is not a mere self-evident and more "clear-eyed" approach to the problem of climate change: it is a "solution" handed over by the very accidental logic that has led us into this dead-end alleyway of global warming. The first worldwide radiation accident generated by the offspring of the Manhattan Project demands the production of an entire national, technomilitary apparatus, which will eventually be hardwired into national-civilian consciousness in the "state of exception" suspension of the juridical. At such a point, a new, axiomatic existence will come into being: life and the military, nature and the war machine, will come to be tightly implicated in a reciprocal grounding.

Force Elements: Control, Command, Communication

In his study of the war machine from Pearl Harbor to Iraq, John W. Dower notes the importance given to bureaucratic methodologies during the early developmental stages of the post–World War II national security state, new "*ideological* attitudes regarding bureaucratic competence and responsibility," and new devotions to increased "productivity" through the improvement of "managerial skills."[37] The overt attention given to developing a strong national security state during and after World War II begins to saturate new levels of intellectual activity. The World War II race to break "enemy codes" required an intense, microscopic rigor too taxing mentally for any single human being to handle. Information needed to be "processed" at greater and greater levels of speed that were beyond the mental scope of any individual human brain. In the field of military theory, the solution created to solve this problem of information processing was called C3: command, control, and communication.

In the military lexicon, *command, control,* and *communication* are terms meant to express the heart of the war machine's administrative functionality—the capacity of the machine to exist as a well-designed, coherent, and powerful force able to supersede all other human sociopolitical forces. The terms contain an essential relation to science but are also meant to reflect how the military stands above the sciences as the higher-power activity that releases the inert or inactive potential of research into productive, usable action. As organizational analytics expanded with the development of computer technologies and information processing and security structures grew in the first decades of the Cold War, C3 expanded to become C3i, adding "information" to the organization motor of command. Military information specialist Thomas P. Coakley describes C3i as follows: "The concept of C3i probably originated in an attempt to apply systems analysis, with its connotations of mathematical precision and efficiency, to command and other functions which directly support command. Take away those things usually thought of as the substance of defense—weapons, ammunition, fuel, logistics, spare parts, buildings, people—and C3i is what remains. Ideally, C3i is what melds the 'stuff' of defense together into an effective fighting machine."[38] This organization of intelligence is more than a simple sign of mere bureaucratic business. C3i names an attempt to discover, in the formal sciences, the fundamental basis for administering a militarized higher-order motor. Its importance matched mammoth energies and efforts that went into the development of SAGE, Arpanet, and the defensive networks that followed: "Both President Carter and President Reagan made improving C3i a top priority, according it, in the words of Senator John Tower, 'equal value with the weapons systems in the competition of dollars and the attention of senior Defense Department management.'"[39] C3i came to be a sign of the human body writ large as a global array, likened to the human body's central nervous-communication system, receiving and processing information and sending orders through the nervous system of widespread chains of command to be processed and acted upon by the body's "muscles." It ties formalized military intelligibility to a global targeting machine and the fastest possible response to an enemy threat: "C3 technology encompasses the capability to acquire, process, and disseminate information across force elements (including international coalition forces). The capability must be reliable, provide secure multilevel access, and be protected from enemy attacks. This will require advances not only in computing hardware and software but in the interconnecting fabric of communications. The goal is seamless and effective integration of capabilities for planning and preemption, integrated force management, and effective employment of sensor-to-shooter system-of-systems."[40]

The terms *command*, *control*, and *communication* arise out of the register of technology; they are not something added onto the technological as an afterthought. In fact, it might be more proper to say that C3i is the very technological structure of the military in that it is the military's apparatus or "equipment" for managing and running its entire set of operations. When examining military manuals and textbooks, one discovers that the essence of the military is not in its hardware or its weapons systems. These are considered by the military to be the physical products or tools created to handle a much more profound dilemma, what we might call the military's primary concern for being:

> Underlying . . . weapons systems and the methods by which they are caused to function, is the fundamental concept that they are simply devices or processes that serve as tools to augment the capabilities of a human being. For example, an electromagnetic sensor may augment the visual and audio sensor of an individual; a weapon enhances the individual's ability to inflict damage; a combat direction system expands the decision making capacity of a person; and so forth. At the functional center, no matter how remote, of every weapon system, combat system, or combat direction system is a human being who is engaged in the employment of that system or grouping of systems.[41]

The concept of the "human" is not theorized here, but we can see that it names essentially a targeting animal. The system of communications, control, and command is the key intelligibility technology that makes possible the extension of the human mind onto the battlefield, directing the mind toward the overall goal of effective targeting. The very possibility of thinking about the human in this biomilitary fashion coincides with a particular relationship to nature and life. The C3i system is this particular structure designed to "augment the cognitive functions of the individual." Communication refers fundamentally to the transmission of information and orders "from the commander to subordinates." Control of the field of reality depends on a dependable and rapid communications network, and C3i systems reflect the technological information imperative that we saw operative in the SAGE defensive shield: "The increasing need for responsive C3i systems is being driven by the rapidity with which weapons can be deployed. . . . [T]his rate has become nearly exponential since the turn of the century." Temporality—shortening reaction times to an incoming weapon once it is initially detected—becomes a target as well of compression. Communication is a mode of human exchange designed to track moving objects; it is an ocular-based method fastened to spatiotemporal coordinates and the fixed reference points that stand out against the movement of (enemy) objects: "Communications: the ability and function of providing necessary liaison to exercise effective command between tactical or strategic units of command."[42]

"Communication theory" as a subject of study did not become a prominent field until the landmark publication of Claude E. Shannon's essay "A Mathematical Theory of Communication" in 1948. These and subsequent works concentrated primarily on the problems of transmitting information within the technical domain of telecommunication, where digital transmission systems needed to make information "measurable" so as to erase "redundancy" in language.[43] De Landa productively telescopes the movement of communication and its importance along the entire chain of technomilitary networks: "Communications technology, 'the step-child of war' as Martin Van Creveld has called it, evolved to the point where telegraph and railroad networks, together with the creation of a loyal citizen army and the imposition of a meritocracy from above, made possible the 'motorization' of the armies without the need to undergo the ordeal of revolutionary turbulence."[44] Communication theory became a means to redirect language in a specific direction toward the goal of entraining a loyal citizen army through a kind of intelligence propulsion system.

This is the essence, for instance, of President Eisenhower's response to the Soviet Union's successful launching of Sputnik, which forewarned America of Russia's possible dominance and exploitation of the region of space through the use of satellite technology. In direct response to Sputnik, Eisenhower created the Advanced Research Projects Agency. Realizing the growing need for "high-level organization within the Department of Defense" in order to "expand the frontiers of technology," he produced what has been called the "single most influential agency in the history of computer development" in the United States.[45] In its current manifestation as the Defense Advanced Research Projects Agency (DARPA), the agency continues to be a major force in the Department of Defense. Its primary concern in the post-9/11 era is the transfer of technology, formerly maintained and managed within the Department of Defense, to the global commons. This constitutes, in the mind of DARPA's deputy director Kaigham J. Gabriel, the current state of geopolitical existence as one in which we are constantly subject to "emerging threats" to national security, pressures always "looming on the horizon." Speaking before the House Armed Services Committee, Gabriel spoke of the inherent danger in the expansion of technology beyond the control of the military industrial complex: "Computing, imaging and communications capabilities that, as recently as 15 years ago, were in the exclusive domain of military systems, are now in the hand of hundreds of millions of people around the world."[46]

Commercialized technology—the lifeblood of today's neoliberalism in the digital age—presents the new threat of "cybersecurity," extending the spatiotemporal coordinates of the traditional battlefield. C3i and its communication theory serve as the higher-order motor that keeps this threat in check.

The avowed goal of the C3i motor is its ability to eradicate insecurities. However, it has the opposite effect, for it cannot exist without constantly uncovering and targeting new threats to reality. This effect appears in the constant demand for "data"—the lifeblood of the war machine in the cyber age. *Data* is the name given to all "unprocessed matter"—a pre-administered form of life akin to the old common fields demonized as "unimproved" by the discourse of enclosure. It is the material from which and out of which the C3i motor begins its movement and becomes an active state of existence: "Data is the raw material from which useful information is made. In isolation, data is meaningless. Information is an ordered or sorted grouping of data that conveys rationality. The purpose of an information system is to process data in aggregation to provide knowledge and intelligence. This information may very well be data for some user on a higher level who may, in turn, process it to produce more comprehensive information and so on, until the final user is served."[47] Though the first use of *datum* dates from Hammond's *Works* in 1646 and is defined generally as that which is "given or granted" and serves "as the basis of reason or calculation," its usage—as a marker of the elements needed to construct calculatory flows—comes to mature and evolve exponentially in the twentieth century, not unlike the rise of gunpowder (data analysis, data bank, data capture, data normalization, data processing, data sheet, data storage, data transmission, data types, data warehouse, etc.). *Data* names the reterritorialization of the unstable movement of material existence into usable and nonusable categories. It is the term not simply for raw materiality, as the official definition would have us believe, but the word for a kind of preparatory action that must occur in the information age before something like real military action can begin. In this sense, the problematizing of data by the security society emphasizes the manner in which the security society does not depend on a "regime of truth" (I will discuss this at length in chapter 4). Rather, the fundamental problem always posed by security institutions is how to tap into the natural and formal order of ecosystems. *Data* becomes a name for the traces of that order. In the military world picture, *data*—as the name for an always potential threat—is the information-evidence equivalent of an ammunition supply.[48]

Paul Virilio has productively shown how C3i had its beginnings in the technological connections being drawn between image capturing and warfare during World War I. The form of C3i operative during World War I developed out of the integration of the combat vehicle and the camera into a new form of optical "weapons system"[49]—the aircraft camera "sight machines" that, in chapter 1, we saw connected to the desire to generate precision aerial photography so as to establish a metaphysics of "automated perception." Automatic electronic search systems began, even at this early stage, to replace the "limitations" of the human eye, driving into existence a form of visualization that could operate on a globalized, geological degree. As Virilio argues, this automatic perception in turn would also begin to generate an "automated interpretation of reality,"[50] for the information accumulated would serve as the database for the command center and its qualified personnel. At the same time, new developments in filmmaking arose out of a similar ballistic background, cinematography itself being developed by the French inventor Jules Janssen who stumbled upon the idea after examining the multichambered handgun invented by Samuel Colt in 1832. Janssen had in mind an "astronomical revolving unit that could take a series of photographs."[51] Virilio reveals how the rifle and the birth of the moving-image camera have their origins in the same technological search for a war-oriented sight machine: "On the basis of this idea, Etienne-Jules Marey then perfected his chrono-photographic rifle [1882], which allowed its user to aim at and photograph an object moving through space."[52] The camera would develop as a military sight machine and a major mode of data accumulation during World War I. In fact, by 1914, aviation's primary purpose in the war effort was aerial reconnaissance, not bombing—as we saw with the development of "flying observation posts" and their cancellation of space and time through the manufacturing of a "spellbinding" "view from above," ostensibly offering a "complete and total picture" of the enemy's territory.[53] The form of the human as an inhabitant was being further erased, supplanted by a form of human being that forms its subjectivity in the metaphysical prospect-view performance of spotting and targeting the enemy on the ground from above.

The analytical machines of the C3i introduce, in essence, a new "geometry" of being that begins to saturate existence in and through its explanatory mode of intelligibility. They are forms of representation (*vorstellung*) that overcode the landscape with a particular, predetermined vision, one akin to the irradiation of the landscape with nuclear toxins: "In Vietnam . . . the Americans were quick to see the importance of rethinking problems of aerial observation. A technological revolution gradually pushed back . . . the old forms of representation [which] disappeared [as] . . . instantaneous, 'real-time' information [took over]. Objects and bodies were forgotten as their physiological traces became accessible to a host of new devices—censors capable of detecting vibrations, sounds and smells, light-enhancing television cameras, infra-red flashes, thermographic pictures that identified objects by their temperature, and so on."[54] A new topological field of vision and representation was being invented, one that destroyed former conceptions of physicality in favor of what was most important: the destruction of "viscosity," of all the materiality of existence that interferes with instantaneous access to the acquisition of the enemy and his or her territory. In this sense, ecosystems undergo a form of regimentation (in the full sense of the term *regimenting*: normalizing, flattening out) that erases any extraneous information. If the old singular and heterogeneous movement of nomadic existence had been replaced by sedentariness in the age of enclosure and military uniformity in the nineteenth century, then the twentieth century brought about—through the hyperactive accumulation of data—the new technology of an "object-shaping geometry" that would come to replace geography itself in the digital age. C3i, in fact, turns geography *into* geometry, making geopolitics a matter of geometric algorithms for the transformation of bodies into mathematical formulae for purposes of data reduction, processing, transfer, and production. Geopolitics will come to rely on this new basis of the geometric in order to make its decisions in the great game of international and environmental security policy. The modern nation-state, accordingly, is a geometric formation that has less to do with the customs in common of imagined communities than it does with the algorithms of military perception.

Clockwork Mechanisms, Entrainment, and Environmentality

De Landa makes an instructive distinction between what he identifies as two fundamentally different processes: self-destructive "feedback loops" and "natural," or what I prefer to call *ecologically singular*, processes that follow the laws of thermodynamics (i.e., they tend toward an equilibrium—a point in which the process minimizes its potential energy, such as we saw in the example of the absolutely heterogeneous gaits of a horse). The Cold War arms race, as I argued, is a process caught within the self-sustaining feedback loop of the Prisoner's Dilemma. Its militarized logic is caught up in the defensive and seemingly unavoidable buildup up of arms by both sides. Despite the fact that each side has generated enough arms to destroy the planet several times over (the well-known "mutually assured destruction" or "MAD" dynamic), the buildup nonetheless continues, generating its own internal, self-spiraling (in)sensibility. Denied of any invitation to pass to the limit of its logic, it offers the subject working within its confines no avenue for delinking himself or herself from the only form of movement and energy it creates as it continually advances in a single direction.

This formalist generation of "man the security machine" had its parallel in the natural sciences. Experiments in "systems analysis" went hand in hand with the search for organic forms of control in nature. New forms of scientific expertise had been sought out as early as World War II for the purposes of discovering "the rules" that govern the natural world. The goal was to unearth Nature's method for extending the security fortress—an early form of Sagarin's "natural security." One such natural world higher-order motor that could speak to the demand for a more inclusive, broad-spectrum influence was found in the discipline of zoology and the action of "entrainment." Entrainment refers to the synchronization of internal rhythms to an external rhythm or cycle in the sense of the sudden coherence of a population of entities—a population of crickets, for instance, shifting from a chaotic movement to an organized chirping in coherence. The idea of entrainment was important to the military because it suggested that there were natural processes, apparently visible in nature itself, whereby unorganized or nomadic populations formed themselves without undue pressure into unifying flows. For the military, this meant that one could find in the various ecosystems of the earth natural flows for developing an esprit de corps in humans. If Nature could entrain its divided bodies to work in unison, then so could the military. Moreover, sociopolitical constitutions of entrainment by the military could be justified as arising directly out of the processes of life itself, which meant that the military could dispense with the need for judiciary rules and regulations, religious moralizing, or any other form of justification for its argumentation. It could, in other words, tap into the natural ground of a state of exception, ruled entirely by necessity. Entrainment became one of the key formative forces influencing the U.S. war machine in the twentieth century. Its level of significance could be compared to the importance of the panopticon in the modern era, as identified by Foucault. Entrainment as a source of behavioral control achieves immense influence when the U.S. national security state begins to petition other nations during the Cold War struggle to contain the threat of expanding communism. At such a point, mass extensions of the fear of the enemy-other—such as xenophobia and jingoism—begin to play an increasingly important role in the organization of a population. De Landa, for instance, proposes that the formation of the nation as an essentially xenophobic entity arose as an answer to the problem of how to maintain a totalized unified field of the C3i structure and an esprit de corps. As the nation began to develop and replace older political forms, the former model of ruler and ruled came to be replaced by an apparatus that would more efficiently ensure that a population would maintain its status as a "reservoir of loyalty"[55] Put differently, when the clash of rulers is replaced by the clash of nations, population becomes an issue and the biopolitical state becomes the stage of militarized politics. Rather than a war between rulers, one goes to war with and against an entire population— and both sides are understood to be two differently entrained organisms. Henceforth the war machine must discover the means to achieve greater forms of devastation: "Battles of annihilation, once the exception, became the rule."[56] Entrainment and the spread of mass destruction become the new imperatives governing the war machine's engagements with environmental security and manipulation.

The military intervention in both the natural and formal sciences that I've only begun to trace here does not simply mean that a war machine dynamic influences and controls what becomes public or that it consciously emphasizes certain forms of technological development. It introduces and enforces "a definite 'style,' a military style that could increasingly subordinate the evolution" of technology and scientific applications "to the needs of the command imperative."[57] The evolution of the projectile wall, aerial photography, nuclear weaponry, data processing, communication methodologies, and entrainment did not, as we have seen, occur in a disinterested fashion. These modes of intelligibility unfolded in a particular direction determined by military needs and the imperatives of security. In searching for the "security foundation" of life, the military had to risk the lives of populations and, with the buildup of arms, the life of the planet itself.

We are thus in the midst of a mode of intelligence that must properly be understood as a self-destructive form of automatic consciousness production. In the mass amounts of information being collected by the self-sustaining, intelligence-gathering machine, only the data that prove to be a threat will be "recognized" and separated out in the never-ending process of logistical compiling. Everything else will be discarded as so much background noise. The importance of the projectile wall of faster computers, able to engage in "number crunching" faster than humanly possible, stems from the need to constitute an apparatus of surveillance that must endlessly seek out new information everywhere and at once—an apparatus that is capable of functioning automatically and at a geological level. This is generally not seen as a specific form of knowledge production. Instead, it is seen as the amplification of intelligence itself, a creation of intelligence in its ideal form, operating at greater speeds, dealing with greater amounts of information, and collating and producing "results" faster than any previously existing organism. In the obsession with "artificial intelligence," we see the crowning of a specifically mechanical, unidirectional form of thought overriding all other possibilities for human intelligence and creativity. No wonder, then, that the various, non-mechanical forms of thinking that occur in other disciplines have come to be viewed as an occasional need for entertainment, if serving anything at all.

These intimate relations—between the sciences, engineering, ecology, and the military—need extensive critical attention. Technological innovations no longer unfold outside the arrays of security and their reterritorializations of ecosystems. As we saw in chapter 2, the very rationale for the biofuel industry, to name one example, has little to do with the search for a sustainable existence. It is governed by a concern for terrorism, the desire to plan for the future political-enclosure securitization of specific geographical regions, the inability of certain agricultural and engineering disciplines to delink themselves from the biofuel problematic, and the recognition that the age of oil will soon come to an end. This security problematic, moreover, does more than influence human thought; it *generates* specific patterns of thought, patterns designed to target a population living through a geological time of turbulence. Environmentality thrives on turbulence, and during times of peace, because its power flows from turbulence itself; it attempts to generate turbulence out of the most mundane moments of existence. As we've seen, the overall effect of sedentary, monomaniacal securitized control produces more insecurity—collapsing all sublime exteriority within a totalizing Accidental field that results in the production of more lethal and widespread accidents—a very special kind of end-of-history peace: installed and "programmed catastrophe."[58] It will be the task of the next chapter to consider some alternatives to this geopolitics of (in)security—to explore how these unproblematized and generally unseen mechanisms set the stage for our access to ecological matters, predetermining the forms in which an event like climate change enters the progressions and extensions of human thought.[59]

4

Genealogies of Military Environmentality II

Environmental Exceptionalism

> "The military isn't waiting while Congress and the general public might be having some debate. They're stepping out as they have on so many other things," Cuttino said. "If there's anybody that's going to be at the forefront of how to save energy, reduce global greenhouse-gas emissions and become more efficient, it's [the military], because it's in their best interest," Cuttino continued. "If they can do it well, it proves to the rest of us that we can do it well . . . We're all going to benefit from what they're doing."
>
> —Phyllis Cuttino, director of global warming, Pew Charitable Trust

> We can't stop short. If we stop now—leaving terror camps, intact and terror states unchecked—our sense of security would be false and temporary. History has called America and our allies to action, and it is both our responsibility and our privilege to fight freedom's fight.
>
> —George W. Bush, "State of the Union Address," January 29, 2002

> The Cold War was a specter, but climate change is inevitable.
>
> —Gordon Sullivan, former U.S. army chief of staff[1]

In the last three chapters, I have explored the development of a particular paradigmatic pattern of thought—a unidirectional, methodological imperative attempting to redefine and govern ecological relations in the twenty-first-century era of climate change. This compulsive and phantasmatic mode of reasoning, which I have been calling *environmentality*, reveals its asphyxiating, self-destructive nature in the closed-loop system of the Accidental. I argued that environmentality appears most emphatically on the ecological register in the empiricist logic of "adaptation," a sociopolitical plan of action that minimizes searches for less destructive ecological relations—especially those that would attempt to consider genuine alternatives to current anthropological–neoliberal environmental relations. In chapter 2, I endeavored to reveal the genealogical lineage informing this fantasy of environmentality: the enclosure movement and the erasure of inhabitancy through the supplanting of custom with common law, which achieved a planetary dimension in the nineteenth and twentieth centuries. In chapter 3, I more directly examined the military dimension of environmentality in terms of nineteenth- and twentieth-century military evolutions and connections to the sciences. In this chapter, I return to what I identified in my introduction as the fourth, or post-9/11, phase of contemporary environmentality.

I begin with a more theoretical consideration of these developments by examining the work of theorists such as Michel Foucault and Giorgio Agamben and New Americanists such as Sacvan Bercovitch, Donald Pease, and William V. Spanos—situating their work specifically in relation to the rise of the security society. Pease and Spanos have traced the history of "American exceptionalism" to the Puritan securitization of the environment and, in relation to the Bush administration's War on Terror, have gained considerable critical leverage by enhancing Agamben's conception of the securitized "state of exception." As I argue in this chapter, the "state of exception"—in which an overwhelming necessity is used as the pretext to suspend the law—generates a political form of action that considers itself to be postpolitical. To put this in the specific terms of my focus in this book, the environmentality expressed by the military in its adoption of climate change as a new mandate is a mode of direct action that presents itself as an apolitical, ethical obligation. In the face of the endless debate surrounding climate change—a debate that sabotages any attempt to stop our destruction of the planet—the military claims it can break through this blockage and *take direct action*. It presents climate change as an *exceptional state* of existence. Hence special "adaptation measures" are authorized. As a staged performance, "direct action" mystifies the ideological nature of these measures: they are presented as an empirical, metaphysical answer to the matter-of-fact dilemma of global warning. This claim of direct action—a perverted form of the Lacanian "act" of breaking through an existing sociosymbolic order—I connect to the strategic deployment of insecurity by the security society.

As I have been arguing throughout this book, the security society cannot function without the strategic and continual deployment of insecurity. This paradoxical structure of security problematically suggests that the security society functions primarily out of the ability to spread fear. As Foucault productively argues, however, the modern era of governmentality functions most efficiently by generating *productive*, not repressive, flows of power. To put this in the insipid language of George W. Bush, the attractiveness of security and military action lies in the positive exercise of "fighting freedom's fight." In order to develop this crucial security schema—the essential relation between "productive" forms of power and the deployment of insecurity—I turn to Foucault's late work on what he also referred to as the "security society." The twentieth century's escalation in security measures, he argues, is made possible by an essential change in the character of freedom. In the age of environmentality, the freedom to choose—to take action against climate change—is turned around so as to affirm all the more the empiricism of necessity. Security, at this moment, reveals itself to be the obscured underside of freedom. The security society's deployment of "freedom performances" leads to the development of what I call "the Accidental state"—a diffuse, asymmetrical structure organized around a fundamental ecological unpredictability generalized beyond any national or continental geographic borders. More and more the earth becomes the ground zero of ecological catastrophe, and in the midst of this cataclysm, our access to freedom from this abyssal milieu is only made possible by accepting the securitizing measures of the war machine. At this precise moment, the war machine takes over the former role of the ecosystem as supporter and sustainer of human life.

Military Environmentalism: The Future of the Ethico-Political "Act"

The national security state of the United States is typically understood as arising out of the Cold War and expanding into its "exceptional state" after 9/11 under the auspices of the George W. Bush administration. There has been little effort on the part of scholars to interrogate the increasing relationship between the security society and the environment. In fact, to my knowledge, only three scholars have directly and extensively drawn connections between the environment and military influences (Elizabeth DeLoughrey, Mike Hill, and Rob Nixon). There is good reason for such a connection to be overlooked. The Bush administration has always denied the fact of climate change. In September 2001, roughly two weeks after 9/11, the Intergovernmental Panel on Climate Change (IPCC) published its third in a series of climate change assessment reports.[2] Barring the Fourth Assessment Report that appeared in 2007 (discussed in chapter 1), this earlier report was perhaps the most clear-headed and shocking statement to date about the extensive and destructive nature of human-produced global warming. Overshadowed by the attacks on the Pentagon and the World Trade Center, and by the George W. Bush administration's absolute denial of global warming, the report, though widely discussed outside the United States and instrumental in the ratification of the Kyoto Protocol, remained a dead letter in American politics. In the eyes of the Bush administration, it had little to do with the "biopolitical settlement" soon to be constituted in line with the fantasies of 9/11 that would come to define the first decade of the twenty-first-century security society.[3] The administration's unleashing of the War on Terror spectacle dominated the media, delimited political consciousness, and bequeathed to Americans their historical role as the true bearers of freedom. It established the "state of exception" as the norm and authorized the American military machine to deploy "shock and awe" tactics to achieve the "spreading of freedom"—conspicuously rendering, in other words, the inherent violence hidden within the brand of freedom engineered by the end-of-history telos of American exceptionalism.[4]

However, as we saw, by the time the IPCC released its fourth report in 2007, the Bush administration's War on Terror narrative had lost much of its efficacy to mobilize the populous in the years preceding and following the Iraq War. Even so, Bush's immobile stance as a denier of climate change meant that the report, not to mention the mounting evidence being published by other scientists, was utterly ignored by the White House. But as I detailed in the introduction, each of the branches of the Department of Defense began to hold conferences and to produce colloquium briefs, strategic communication plans, reports, and complete dossiers on climate change. While the Bush administration was still emphasizing threats of terrorism, the Pentagon was attempting to tightly weave planetary environmental crises into this post-9/11 Cold War replacement. To again quote Admiral T. Joseph Lopez, former commander-in-chief of the U.S. Naval Forces, "climate change" would come to "provide the conditions" that made it possible to "extend the war on terror."[5]

This expansion of the security society and the military machine into the unlikely arena of environmentalism reflects, in part, the ideological shift to the permanent state of war that began in the wake of World War II with the signing of the National Security Act of 1947 and the identification of the Soviet Union as the United States' chief enemy-other. The permanent state of war constitutes one of the early signs of the "state of exception" as identified by Agamben while also putting into play the idea of "American exceptionalism" identified by New Americanist critics as a founding ontology of the U.S. nation-state. As numerous scholars such as Sacvan Bercovitch, Donald Pease, William V. Spanos, and others have argued, the fantasy of American exceptionalism extends back much further than the occasion of 9/11. The establishment of the state of exception as the norm during the Bush administration (namely, the authorization of preemptive war, the termination of civil liberties and the transformation of panoptic security surveillance into law with the Homeland Security Act, and the establishment of "stateless citizens" outside the law in "extraordinary rendition" sites of Abu Ghraib and Guantánamo Bay) reflects an older idea of exception, operative even before the inauguration of the national security state in 1947 in America's Puritan foundations. The Bush administration's actions "repeated . . . on a global scale and in secular terms the essential ontological, cultural, and political principles that were intrinsic to the American Puritans' founding of the 'New World'. . . . [and] their representation of their status as a 'Chosen people.'"[6] This "deep 'exceptionalist' structure . . . of American national identity," a divinely ordained "errand in the wilderness," metamorphosed into the "secular ideology of manifest destiny in the westward expansionist nineteenth century and into the global truth discourse of America in the aftermath of World War II."[7]

Donald Pease articulates the specifics of the national security society in terms of what he refers to as the "internal externality of the exception":

> The declaration through which the state inaugurated [the Cold War] materialized a site whose surplus violence rendered it inassimilable to the terms of the national order. In conducting the cold war, the state was neither within the order nor outside the order. The state situated itself within the order that it protected but it

occupied the position of internal externality of the exception. For in order to defend the order it also repressed, the state was first required to declare itself an exception to the order it regulated. The State of Exception is marked by absolute independence from any juridical control and any reference to the normal political order. It is empowered to suspend the articles of the Constitution protective of personal liberty, freedom of speech and assembly, and the inviolability of the home and postal and telephone privacy. In taking up the site of the exception, the cold war state instituted a permanent alternative to the normative order that it called the National Security State.[8]

Pease articulates the paradox of the superstructural nature of the security state—existing in a suspended zone that is neither within nor without the political order. Put differently, the justification for the naturalization of a state of exception arises from the idea that in the face of an extreme threat, security serves as the self-evident solution to that threat. This is the logic lying behind such paradoxical patriotic statements such as "freedom is not free" that appeared in the wake of 9/11. However, in terms of the ecological paradigm of environmentality that I have been exploring here, the state of exception as identified by Pease has begun to shift into a new stage of development. It not only functions as the norm but also now declares itself to be the true nature of existence. Justifying its existence and supremacy in the fantasy of an ontological guarantee, it presents itself as the true order of life itself: *natural security*.

Natural security depends on a constant state of crisis, a survivalist logic that depends on a certain inherited ideological observation of nature. Prior to this stage of the development of the security society's exceptional state and to Pease's characterization of the exception's internal externality, Sacvan Bercovitch identified the exceptionalist state in terms of "the American jeremiad," a covenantal community organized around a threatening wilderness that, in its representational deployments of nature as an unknown and dangerous frontier, generated a constant state of anxiety that turned "crisis into [a] social norm": "Its function was to create a climate of anxiety that helped release the restless 'progressive' energies required for the success of the venture. . . . Like all 'traditionalist' forms of ritual, it uses fear and trembling to teach acceptance of fixed norms. But the American Puritan jeremiad went much further. It made anxiety its end as well as its means. Crisis was the social norm it sought to inculcate."[9] We see this same element of crisis transfigured from the Puritan wilderness discourse to the twenty-first-century occasion of natural security, which tenders nature itself as the self-evident proof for the expansion of security and exceptionalism to contemporary environmental arenas. In terms of the military connections I wish to draw in this chapter, I trace the "security jeremiad" back to the earliest formations of the American military industrial complex—specifically, to the origins of the relation between Enlightenment science, technological progress, and the security state's normalization of war as the state of exception. These specifically American and military origins, as we will see, constitute a more pronounced form of the enclosure paradigm of existence I analyzed in chapter 2.

One additional feature should be kept in mind. As in my previous exploration of the erasure of inhabitancy as a viable form of subjectivity, the point of this analysis is not to identify the fantasy of an exceptional state to be "false," which would require, in turn, the establishment of a "true" position that would require less criticism. Rather, my method is to examine the principles of rationalization that link together different practices that are, because of these very principles, found to be convincing. For instance, up until the seventeenth and eighteenth centuries, inhabitancy was found to be a convincing form of human existence in the West. After the creation of a new judicial system (common law), the individual and the entrepreneur replaced that subjectivity. Each is "true" according to its different principle of rationalization. Each establishes, to speak like Foucault, its own "regimes of truth": "The point of all these investigations concerning madness, disease, delinquency, sexuality, and what I am talking about now, is to show how the coupling of a set of practices and a regime of truth form an apparatus (*dispositive*) of knowledge-power that effectively marks out in reality that which does not exist and legitimately submits it to the division between true and false."[10] It is worth pausing on this powerful conceptualization of truth regimes in terms of the maneuvers of the military in the fields of natural and formal sciences that I examined in the previous chapter and in terms of Foucault's increasing interest in mechanisms of security in the last years of his life. Foucault emphasizes the division between the true and the false as the primary markers of the government of humanity in the latter half of his career. Apparatuses of knowledge-power, he argues, are justified in terms of their ability to shift and define the parameters of what counts as "true" and what should be avoided as "false." A regime thus endows itself with power (or more accurately, takes on a flow of power) by characterizing itself as acting on the side of truth. The basis of the regime is thus the fantasy of its self-referential establishment of its flow of knowledge-power as "true." This fantasy is its ontological guarantee, for it is "in reality that which does not exist." In other words, the regime makes itself appear as a positive existence that has always been present but simply not fully articulated or "discovered." The fantasy is thus not an illusion, as I described before in chapter 1, nor is it an error, against which one would then posit an unquestionable truth. Rather the fantasy, or "the regime of truth," "makes something that does not exist able to become something. It is not an illusion since it is precisely a set of practices, real practices, which establish it and thus imperiously marks it out in reality" (19).

More interesting for my purposes here is the way that Foucault historically situates the rise of the regimes of truth. At the opening of his lectures on the birth of biopolitics in 1978 and 1979, he makes a distinction between three different epochs of control in the modern age. (It should be remembered that this late lecture was after he began to emphasize the concepts of "biopower" and "governmentality" and shift away from discussions of disciplinary societies to an analysis of the "security society.") The first epoch spans the sixteenth century to the mid-eighteenth century—roughly the time when common law was struggling to replace the heterogeneity of custom in England. Foucault also has England in mind when he characterizes this epoch as sovereign and feudal in nature, but it is also, importantly, a form of statehood that fails to achieve an extensive apparatus of control. He defines the epoch as having a range of practices (taxes, customs charges, manufacture regulations, etc.), and this list tells us that he is essentially talking about custom law. He points out that these practices did not stem from a transcendent armature: "All these practices were certainly reflected on, but on the basis of different events and principles of rationalization" (18). It is not until the latter half of the eighteenth century, according to Foucault, that a generalized rationality came to exist: "From the middle of the eighteenth century it becomes possible to establish a reasoned, reflective coherence between these different practices . . . a coherence established by intelligent mechanisms which link together these different practices and their effects, and which consequently allows one to judge all these practices as good or bad, not in terms of a law or a moral principle, but in terms of propositions subject to the division between true and false" (18). Though he does not name it, the "reasoned coherence" of which Foucault speaks is common law.

After describing these changes, Foucault begins to define the stage of reasoned coherence as being made possible in and through the division between the true and the false. This division, he argues, is the dimension in which today's form of governmentality functions. The age of feudalism was marked, he argues, by its emphasis on sovereign power, on bringing government to its "maximum strength." The problem of government in the dimension of regimes of truth, however, asks a fundamentally different question: "Am I governing at the border between the too much and the too little, between the maximum and the minimum fixed for me by the nature of things—I mean, but the necessities intrinsic to the operations of government? The emergence of this regime of truth as the principle of the self-limitation of government is the object I would like to deal with" (19). The limitations of governing stem, in part, from their new liberal regimes (the liberality of the market, of the individual, etc.). But this liberal freedom is not a given; it is produced and organized by the state and the state's acts of dividing the true and the false. It is on the side of too much liberal activity (the false) that the manufacturing of freedom must resort to a necessary "principle of calculation." This principle of calculation is what Foucault identifies as "security." Strategies of security, says Foucault, are "liberalism's other face" and "its very condition" of existence: "The game of freedom and security is at the very heart of this new governmental reason. . . . [T]he economy of power peculiar to liberalism [is] internally sustained . . . by this interplay of freedom and security" (65).

In this chapter, I want to posit the suggestion that environmentality is not, or not primarily, about the division between the true and the false. Rather, as security begins to gather greater weight, authority, sway, legitimacy, and, perhaps most importantly, validity across a general national population, the importance of the division between the true and the false subsides, and a new form of "the natural," coupled to the idea of "necessity," begins to take over. Arguing, for instance, that nature is essentially about security is not an argument attempting to put forth a fundamental truth (despite the fact that notions of truth may help this argument out). The argument that nature's ontological ground is the fight for one's security and the securitization of one's territory is essentially a *militarized* argument that interprets the basis of life as warfare and survival. It is not a matter of right or wrong at this primary level; it is a matter of continuing to exist or of losing one's existence. Thus it is more proper to refer to environmentality as a regime that attempts to unleash upon socio-political reality a fundamentally military interpretation of "animalistic nature"—that is, a knowledge of nature that understands that defending one's territory, or killing in order to survive, is neither true nor false. Such activities are understood to be an empirical fact of existence. Once again, it is important to understand that this empirical facticity is not an error, nor an unquestionable core of reality, but a fantasy that enables something that does not exist (all nature is always and only about security) to become the energy that drives the discourse of environmentality and the apparatus of the security society that seeks to defend that discourse.

It should be remembered that in *State of Exception*, Agamben locates the origin of the state of exception in the modern era at the commence-ment of military power's extension into the civil arena, specifically with Napoleon's decree of 1811, which introduced an *état de siège* (state of siege) capable of being declared even when a city or a people were not being threatened. The *état de siège* differs from the *état de paix* (the state of peace in which military and civil authorities are completely separate) and the *état de guerre* (an actual state of war in which civil authority acts in con-cert with military authority). The *état de siège*, Agamben emphasizes, was originally associated with actions taken during a time of war, but it even-tually disconnected itself from any actual legally declared state of war and took on a life of its own: "The subsequent history of the state of siege is the history of its gradual emancipation from the wartime situation to which it was originally bound in order to be used as an extraordinary police mea-sure to cope with internal sedition and disorder, *thus changing from a real, or military, state of siege to a fictitious, or political one.*"[11] In the latter half of the nineteenth century, the nonwartime *état de siège* came to be coupled with the suspension of the constitution: "Although the paradigm is, on the one hand (in the state of siege) the extension of military authority's wartime powers into the civil sphere, and on the other a suspension of the constitution (or of those constitutional norms that protect individual liberties), in time the two models end up merging into a single juridical phenomenon that we call the *state of exception.*"[12] This merger produces what Agamben identifies as a "kenomatic state, an emptiness of law," into which the sovereign steps as the person who suspends the law. At such a point, "emergency becomes the rule, and the very distinction between peace and war . . . becomes impossible."[13]

There are two important aspects of Agamben's research that need to be underlined here. First, Agamben identifies the state of exception as "fictional"—that is, as a political construction that exists at the limit of the legal system (or as an "internal externality," as Pease articulates it). In this sense, it is a fantasy: real but having no grounding in an ontological guarantee. To refer back to Jacqueline Rose's articulation of the relationship between the state and the structure of the national (citizenry) psychological fantasy, the nation-state fantasy names the dominant structure of desire and not a simple notion of mystification that would readily vanish in the face of revealed "truths." The state of exception is therefore empty of (customary) legal rules and operations. But it is also empty in a different, "kenomatic" sense:

> The expression *full powers* (*plein pouvoirs*), which is sometimes used to characterize the state of exception, refers to the expansion of the powers of the governments, and in particular the conferral on the executive of the power to issue decrees having the force of law. . . . The presupposition here is that the state or exception *entails a return to an original, pleromatic state* in which the distinction among the different powers (legislative, executive, etc.) has not yet been produced. As we will see, the state of exception constitutes rather a kenomatic state, an emptiness of law, and the idea of an originary indistinction and fullness of power must be considered a legal mythologeme *analogous to the idea of a state of nature.*[14]

The pleromatic is the realm of the gods, the metaphysical, Platonic ideal realm of fullness. But as Agamben shows, plenitude and full justification of the sovereign declaration of the state of exception is an artificial pronouncement having no firm basis in any ontological certainty. Key to my analysis here is that this lack of fullness—the emptiness of the kenomatic state—attempts to secure its authority by attaching itself to the idea of an original state of nature. This is also the origin of the nexus between the emptiness of the state of exception and an empiricist philosophy. *If the state of exception can be shown to be the original state of nature, then the study of nature* (which has, of course, a long history of being fastened to empiricism) *can support the expansion of the security society's exceptional enforcements.* Thus, in an idea such as "natural security," the kenomatic and the empirical work together seamlessly. In this way, and as we will see in more detail in this chapter, the fundamental problem always posed by security institutions is one not of truth but of how to tap into the natural and formal order of ecosystems—into, that is, environmentality.

Military Environmentalism and the "Organic Necessity" of Technological Progress

The exalted idea of technological progress, as many environmentalists and ecocritics have pointed out (Commoner, Rodman, Capra, Merchant), has from its beginnings in modernity been distinguished by a mode of disinterested inquiry indifferent to the relational nature of ecosystems. However, not widely acknowledged or examined is the inherent connection of this disinterested inquiry to the security society idea of a presumably organic, inexorable movement of contemplation. As Foucault defines the identifier "security society" in his later work, it is a form of governmentality that functions differently from Foucault's previous identifier of modern Western sociopolitical reality, the "disciplinary society." The security society names the state of existence when war becomes the dominant form of politics—that is, when the state of exception reigns and consequently requires the war machine to be folded into daily life.

As we saw in the previous chapter, the security society would be buttressed in the twentieth century by certain developments in technological progress—developments that were marked by the incorporation of a unidirectional strategic and logistical approach leading to the creation of the military industrial complex. Disciplines—namely, certain formations of physics and biology and later information technology, computer science, and game theory—came to grow in line with a pattern of experimentation that was analytical, linear, and strategic in nature (as opposed to the systemic and holistic approach favored by environmentalists). Recall the difference between classical and molecular biology and the erasure of the complexity of the whole cell by molecular biology's targeting of discrete elements in a cell at the expense of their interaction.[15] This target-oriented, linear analytic correlates with the continuing faith in capitalism as the only economic game in town. In this sense, the analytic is of a piece with the self-limitation of government that Foucault identities as the principle of governmentality. Governmentality, as we saw, functions by manufacturing as much "freedom" as possible; by discovering and implementing "the necessities intrinsic to the operations of government," it allows neoliberalism a free reign.[16] Even during the Climate Change War

Game, for instance (discussed in chapter 1), the participants held firm in their commitments to continuing neoliberal progress, despite the fact that the game was about the most extensive environmental threat in modern human history: "Throughout the game . . . teams [United States, China, India, etc.] never wavered in their drive to balance any agreement with economic growth."[17] This relentless passion for maintaining the freedom of economic growth is indicative of the continuing ruthless pursuit of not only financial gain on the economic register of being but also the control of resources (and the territories associated with resources) that defines national security interests.[18] It is primarily, even in the face of catastrophe, *the continuation of a form of freedom that generates catastrophe.*

The alliance between the military and the corporate world of unrelenting economic overdevelopment and overconsumption is nothing new. However, what is not often acknowledged is the relation of this linear analytic to the problem of diminishing resources and returns. For instance, it is not difficult to realize that by the end of the twenty-first century, current conceptions of the economic will no longer be applicable. The very concept of "economic growth" and the models generated by its demand are already becoming outdated. In previous centuries, and even as late as 1973 (i.e., before the Arab oil crisis and the advent of "energy security" examined in chapter 2), such growth was based on the naïve view that resources would always be available and the view that the earth is a mere resource for the anthropological consumption machine. As we enter the age of resource wars, we shift into a mode of existence that will be underwritten by the knowledge that the resources we covet will soon come to an end. By the end of the twenty-first century, as projection models indicate, the resources that currently define human survival will no longer be available. Oil production is expected to peak sometime between 2020 and 2040 and then decline.[19] Crops are projected to fail more often, even with agronomists' efforts to design new varieties of staple crops like rice and wheat.[20] The "economic" as defined by neoliberalism as a free machine of "growth" will become increasingly compromised and problematic. Despite the search for alternative sources of energy, its central mechanism—production for accumulation of wealth and increased consumption—is at odds with resources, which are always assumed to be available without reserve. The fantasy of new forms of "green technology" and sustainability as being either unlimited or at least far less limited than oil and gas is symptomatic of this assumption. This fantasy of a form of production and consumption without consequences is the mechanism that makes an idea like the free market possible. And it is the flip side of the deployment of necessity. The neoconservative green-friendly stance of "delinking America from foreign oil" is also based on this fantasy of finding and installing in the social system a point of "no consequences" from which all future reality can be manufactured. But because this socioeconomic foundation of "no consequences" is a fantasy, the economic must constantly conceal this illogical limit of its existence.[21] The movement of the economic will consequently be defined as the movement of exhaustion, of a mode of production oriented to the telos of depletion. The Accidental paradigm has already begun to deploy the idea of depletion in order to maintain the "crisis mode" necessary to the successful continual functioning of the security society. More and more, depletion, coupled with the ecological destruction it generates, will power the motor of (anti)development and serve as the captivation mechanism that disinhibits any relation to an exterior alternative to this paradigm.

To return for another moment to the example of the Climate Change War Game, the premise for engaging in the game becomes clearer when we begin to consider the depletion component of the Accidental paradigm. A key rationale working against sustainable innovation stems from the way in which security almost exclusively focuses on "near-term" narratives of insecurity: "With a few exceptions, we took technology projections from the 2007 IPCC report, which some of the scientists in the consortium regarded as too conservative but which demonstrate the near-term limits of relying on technical solutions to climate change"; "The consortium . . . hoped that a major outcome of this game would be for players to leave with a sense that action in the near term is necessary and possible"; "We based energy projected . . . [in] the scenario [on] the International Energy Agency's World Energy Outlook, which sticks to current and highly likely near-term policy, supply, and demand trends more closely than do many other projections."[22] New "technological" solutions to the problem—which are always long term in their implementation—are understood to take away from the immediate threats to national security. If the immediate issues of security are not addressed fully, then all future security crumbles. This logic, presented as plain and disinterested, reflects the self-strangulating dynamic of the closed-loop structure of environmentality. It seeks to release the full potential of climate change, exploding nature as a great destructive force that may erupt at any moment and making it necessary for us to be constantly on our guard and be "realistic" about what will happen not only to our loss of resources and shifting geographies of agricultural production but to the threat to national borders when climate change refugees begin their forced migrations. Pushing sustainability (now referred to more often as "mitigation") to a secondary status, if allowed on the table at all, manifests the logic of redirecting our attention toward the next impending accident. Even making it a secondary concern may not ensure that it will ever take precedence, since adaptation will always already be a more pressing concern. In this sense, the Climate Change War Game repeats, move for move, the strategies deemed necessary by the Prisoner's Dilemma.

This self-destructive pattern of existence constitutes an escalation of the kind of destruction of transcendence in the Cold War scenario of nuclear testing and its resultant planetary contamination that we saw in the previous chapter. *Unlike* the Cold War experiment, however, this new attempt to transfer our environmental and economic future to the dictates of the security society manifests the full-scale potential of the Accidental. Prior to the age of environmentality, the Accidental as an effective paradigm was actuated "by accident," so to speak; now, the Accidental has become a fully "objective" structure of existence—that is, systemic and autonomous. In the discourse of environmentality, *environmental breakdown serves as the very motor of the security analytic and replaces humanity's concern for the environment.* The environment as essentially an entity breaking down becomes the new universal truth of humanity's existence in the world, essentially hardwiring conflict and adaptation to the civilian brain. From this position, it is a small step for the security society to install in the minds of a civilian population the idea that the environment is humanity's ultimate enemy. In this phantasmatic imaginary, the preservation of the planet starts to disappear from human consciousness in favor of walled and viciously defended spaces, all in the name of a pragmatics totally alienated from its own assumptions—a pragmatics unrelentingly extinguished of any other economic, political, or cultural conception of existence.

As we saw in chapter 1, the Center for a New American Security (CNAS) war game exercise is part of a chain of events occurring in the war machine, one of the first in the development of a "coordinated strategic communication plan"—a repetition at the ecological level of the old SAGE communication network—marking the expansion of the national security state into the arena of environmental activism.[23] This is reflected most clearly on the U.S. Army War College's "The National Security Implications of Global Climate Change" conference brief. Like the worst-case scenario of the Climate Change War Game and like the SAGE early warning system, the brief stresses the need to anticipate the greatest catastrophic change: "The entire range of plausible threats needs to be delineated, then analyzed and early warning criteria established."[24] Unlike the war game, the U.S. Army War College's Strategic Studies Institute argues for global cooperation but suggests that such cooperation is not yet available: "Climate change will require multinational, multi-agency cooperation on a scale heretofore unimaginable." Cooperation, however, is to be governed by the linear analytic of the Accidental paradigm: despite the suggestion of such multinational cooperation, the brief makes it clear that a climate change early warning system demands a "catastrophic vision," a mental picture that will effectively reduce all other courses of action to "one of national survival." Summaries of the U.S. Army War College conference presentations indicate that many participants found the then newly published IPCC 2007 report to be too moderate in its predictions. The development of a terminology for establishing a clear security discourse was foregrounded, and the final presentation emphasized the need to keep this "discourse at the national security level rather than the disaster relief level." It was suggested that this emphasis be extended to other state security structures and to become a key focus of the National Security Act of 2010.[25]

In this discursive explosion of military ecological interest, we find the division between security and insecurity and the pragmatics of "necessity" assuming an emphatic role over the division between the true and the false. National defense structures are presented as compromised by the essentially volatile nature of climate change (scientific uncertainty about specific sea level rises, world temperature increases, polar ice cap stability, hurricanes, etc.). The logic of security and its nationalized systems finds its justification in and through the incorporation of a certain form of insecurity in actualizing its existence and its proper functionality: "As military leaders, we know we cannot wait for certainty."[26] In the ontology of environmentality, the act of decision functions by banking on an artificial future deployment of a perverse absolute certainty (ecological catastrophe, failed crops, massive displacement, political unrest, etc.) or what amounts to the same thing: the constant deployment in the present of uncertainty. The certainty of catastrophic collapse in the future exists side by side with the uncertainty of that knowledge in the present. This enables the security specialist to annex disagreement (in both the traditional conception of that word and in the philosophical sense given to it by Rancière) from the domain of the political. The military authority employs the policing idea of "risk" at the expense of the political transformation of the field of possibility so as to justify "action now":

> This approach [ending the discussion and acting now] shows how a military leader's perspective often differs from the perspectives of scientists, policymakers, or the media. Military leaders see a range of estimates and tend not to see it as a stark disagreement, but as evidence of varying degrees of risk. They don't see the range of possibilities as justification for inaction. Risk is at the heart of their job: they assess and manage the many risks to America's security. Climate change, from a Military Advisory Board's perspective, presents significant risks to America's national security.[27]

We thus find the most conservative, policing organization standing on the side of environmentalists that have been attempting to convince the public to take climate change and other ecological problems seriously. It is time to stop questioning whether or not climate change is real: "Debate must stop," says former U.S. Army chief of staff Gordon Sullivan, "and action must begin."[28]

The mandate to "end debate" constitutes itself across each of the Department of Defense's agencies of national security and subordinate departments of armed forces. As Rear Admiral David Titley, head of the Navy's Task Force on Climate Change, phrases it, "Are we going to wait for perfect data? No. Not only the Department of Defense but any successful organization doesn't wait for perfection."[29] It is worth theorizing the deployment of the idea of perfection here. In the Accidental state, certainty can manifest itself contradictorily in its opposite form of "uncertainty." Uncertainty—the unknown threat of the coming accident—becomes the basis for a new and unswerving will-to-power. Uncertainty—or better yet, the Real of environmental collapse, what the symbolic order of the enclosure movement and the economy of capitalism could *not* symbolize and needed to exclude in order to establish their domain—is now incorporated into the system not as its internal contradiction but as its primary source of its energy. Its unwanted and unheeded offspring has become its raison d'être. The security society scorns the Platonic conception of perfection while paradoxically holding onto the idea of a future certain truth (the ultimate accident will happen) that it wants to ostensibly avoid but also needs to "occur" as an unavoidable approaching event. In temporal terms, the present is thus structured in a foreclosure of the openness given by future's nonpresence by a future colonized by the imaginary presence of catastrophe. The future is thus by definition the accident-to-come, not a *future anterior* constitution of the present after the fact; it is a cessation of movement in the present orchestrated not by a democracy-to-come but by the disaster-to-come.

This methodology of "disaster forecasting" is reflected in a number of documents from the Department of Defense and the institutes associated with it. In 2010 the Scripps Institute of Oceanography opened the Center for Environment and National Security, an institution designed to address the national security implications of environmental change. Its prime directive is to help scientists "meet the forecasting needs of military, diplomatic communities."[30] The "military" and "diplomatic" referents are clarified in their relation to the future theater of global warming: the new center will "help the national security and foreign policy communities prepare for the potential societal effects of environmental change—from food shortages to water wars to mass migrations and civil unrest."[31] This generation of a catastrophic narrative is a double-edged sword. For decades, left-oriented and liberal environmentalists struggled to convince the mediatized public and policymakers on Capitol Hill that we were headed toward extreme ecological devastation, and rightly so. Now, as Phyllis Cuttino argues, the military is at the "forefront" of this battle, pushing us to "save energy, reduce global greenhouse-gas emissions and become more efficient."

The problem with the "forefront" of this political battle is its emphasis on the Accidental paradigm by thrusting the future disaster into the present. Here "catastrophe" names a scare tactic designed to distract our attention away from alternative modes of existence. The specific adoption of the "catastrophe" trope also turns our attention away from substantive analysis: "National security experts said climate forecasters often focus on averages, or the most likely scenario, without determining the probability of an extreme climate shift."[32] The push to improve scientific climate change models forces knowledge production about the environment in a direction that emphasizes the total exposure and control of the Accidental. It is a joint venture that includes offices of the Navy, the Air Force, the National Oceanic and Atmospheric Administration, White House science advisers, and a host of others, and it centers on the production of "next-generation climate models" that "incorporate knowledge of the social sciences, agriculture, and marine ecosystems."[33] Herein lies the most seductive aspect of the military subject position: the actor that cuts through all the red tape and takes a strong position of action is assumed to take the ethical position, precisely because he or she sees directly through to the "higher Necessity"[34] lurking behind the senseless contingency of both reality and the empty democratic dialogue that seeks to "debate" the reality of climate change.

The idea of a higher necessity has no credibility without the deployment of disaster. In the face of disaster, necessity trumps all rules and regulations; that is, it trumps the law. Agamben references this relation at the opening of *State of Exception*: "*necessitas legem non habet* [necessity has no law]."[35] It is in this sense that necessity and the state of exception can have no juridical form. We can sharpen this idea of the organizational power of the deployment of a disaster-to-come by building on Agamben's analysis of nature as generative of an original "force of law" (in this syntagma, Agamben writes "law" crossed out).[36] Nature, in being prelaw—the founding accident of existence—is reterritorialized in the sociopolitical domain as the raw materiality from which law and legislation arise as a reactive force, making it possible for order to exist. If the state of exception, overseen by the executive, is the first order that will eventually make the second order of freedom possible (the sovereign suspends freedom in order to save it), then in the totalitarian form of sovereign dictatorship that Agamben presents as the principle or naturalized paradigm of twentieth- and twenty-first-century governments, it is the threat of another nation or state or rouge actor (a terrorist group) that is "causing" the necessary suspension of liberties. But in the age of environmentality, the threat that requires an exceptional, Accidental state is Nature as pure force, as the essential instability that founds power. It is because of *this* Nature that we are required to paradoxically suspend freedom so that the original paradigmatic foundation (a stabilizing force) can be laid, thus paving the way for the law and the future legislative acts that will manufacture the forms of freedom that will reinforce the reigning paradigm to be born. Following Agamben's crossing out of the word "law," we can properly understand the antifoundational force of environmentality—that which offers modern Man the ability to build a reality on the fabrication of necessity, or the "force of ecos" (crossed out). The force of ecos thus refers to a reduction of the complexity of the environment (its various systems, its biodiversity, its complex flows, its enthropic nature, etc.) to a single violent force, one that threatens all forms of freedom and the very idea of freedom itself. For this reason, the representative speaking on behalf of the Accidental state can always say that he or she is always acting out of absolute purity, for the necessity is always represented as being *before* representation (discursive constructions, representative governments, laws and regulations, political agendas, personal interests, etc.).

Herein lies the difference between this military concern for the environment and the concern of the environmentalist wishing to make climate change a vital political issue. On the surface, the entreaty of military leaders reflects exactly the kind of commitment long sought by environmentalists, a commitment that breaks through the endless liberal democratic debate surrounding the issue of climate change and the barriers such as "Climategate" thrown up by pseudoscientists and lobbyists backed by corporations that continue to support hydrocarbon energy sources as the primary method of production.[37] Gordon Sullivan's "debate must end" stance offers compelling relief to these mystifying and crippling forces. He speaks in such a way as to move beyond the fundamental deadlock of debate and even references the same concept of "perfection" cited by Rear Admiral Titley: "We seem to be standing by, and, frankly, asking for perfectness in science. . . . People are saying they want to be convinced, perfectly. They want to know the climate science projections with 100 percent certainty. . . . We never have 100 percent certainty. . . . If you wait until you have 100 percent certainty, something bad is going to happen on the battlefield."[38] In a different context, these remarks could be taken as an explanation of a scientific method, which is also based on an understanding of a certain level of uncertainty in any results and the conclusions derived from results.

The difference between these positions lies in their opposing relationship to decidability. For the scientist, making a decision in the face of a certain amount of uncertainty means that one keeps his or her findings and conclusions—and perhaps even his or her very methodology—open to question. The choice taken by the military leader, on the other hand, puts a different kind of decision making into operation: the decision operates more along the lines of a verdict given in a courtroom. It acts like the sentence of a judge. The nexus of the relationship between (a)political or nonpolitical (in the sense of not being clouded by a limiting political agenda) action and pragmatic empiricism lies in this decision sentencing. The strength of the kind of decision making adopted by the military lies in slicing through politics as usual, which is habitually seen as crippling any productive action that might lead toward a real change of affairs. However, this apparently (a)political[39] action grounds itself in the kenomatic politics of necessity *generated by the discourse of security*, in which government control ("politics") is understood to be in need of restraint. In this kenomatic state, politics is always "too much" and too restrictive of the kind of genuine action necessary to handle empirical reality.

Unlike the endless and innocuous deliberation about the "reality" of climate change that thoroughly engulfs the register of the mediatized public, environmental efforts, the government, and the paralyzed political state of all three, we see in the military commander's "no-nonsense rhetoric" the core of today's environmentality. In other words, this large-scale enfolding of environmental concerns into the military machine has the ability to cut straight through the endless postmodern, "tolerant" chain of signification (the constant but empty engagement with differential points of view). It is not a matter of being "true" or "false" but of setting in motion the machine of an organic empirical reality. As such, this rhetoric is able to co-opt, in a reductive fashion, the Real (in Žižek's sense of the term)[40] of liberal democracy—that is, the ability of a subject or group to assume an overtly political mandate, a larger (populist) cause, *without being demonized* as *political*. This ability of the military to supposedly touch upon the Real of democratic American and global neoliberalism carries the potential to galvanize the population without being turned into what such political movements normally appear to be: the external enemy that must always be avoided (the enemy that "guarantees Society's consistency")— the "socialist," the "tree hugger," and so on. This "political which is not political" is of a piece with such scientific endeavors to uncover the ontological essence of the natural world as entrainment, data processing, and the Prisoner's Dilemma.

This "wisdom" is also marked by the ability to tap into the strategic importance of "speed"—which, as we saw in the last chapter, begins to carry increasing importance in relation to the postpolitical security society "act." Military and security experts give weight to their position on the democratic playing field by identifying one of their key strengths as the ability to make decisions without having to waste precious time waiting for the bureaucracy to produce a conclusive report. In this sense, they set themselves apart from the old nation-state, which lags behind the Accidental state that the military helps bring into existence: "While political leaders on Capitol Hill seek definitive answers about how quickly the world's climate will change, military and national security experts say they're used to making decisions with limited information."[41] This capacity to "get to the bottom of the problem"—to touch directly on the Real and act quickly—reflects the joining of a raging form of speed (a single and thus more uninhibited flow or existence) to the policing-politics (Rancière) that assumes it has "accounted" for everything in existence.

This military engagement with the Real therefore brings about a perversion of the authentic ethico-political act as defined by Rancière. In more than once sense, Phyllis Cuttino of the Pew Charitable Trust is spot on: the military is at the forefront of the environmental fight, working on all levels to achieve what scientists and activists have sought for decades now (of course, the success of all these efforts is still a question). Indeed, as the sudden and increasing production of these and many similar military reports suggests, the military is genuinely performing the authentic radical act of directly assuming responsibility and taking action on behalf of the environment—breaking through the overload of ideological representations that act as a congestion to a concrete act that would, presumably, change the playing field that controls all ideological points of view. However, for an act to be truly genuine, it must be more than a movement that resolves a series of problems existing within a determined field of action; it must enact the "more radical gesture of subverting the very structuring principle of this field."[42] Or to speak like Rancière, the genuine act must name a wrong and put into play—against the naturalized play of accepted parts/identities/activities/representations—the "part of no part." And the part of no part that environmental activists, ecocritics, and the military share is a commitment to break through the mystifying cloud of debate so as to represent the truth of climate change, to name it as a wrong that has been committed, and to take action to fully confront the event. (It should be emphasized, however, that there are other "parts" not included in the existing field of policing-politics and that the question of the "part of no part" becomes more complicated when we begin to address constituencies such as the Articulation of Indigenous People of the Amazon or the

MST—each occupying different "nonparts" than environmental activists from the United States.) If the military does succeed in "turning the tide" toward a new American form of environmentalism, the twenty-first century may usher in a new inauguration, a new genuine "event" of history, to speak like Heidegger—a history in which environmentalists will need to play a "part" in accordance with the new practices of military ecologies.

Decision and Necessity

Another way to sharpen the difference between the military's and the environmentalists' triggering of the act is to consider their close but very different relations to uncertainty, to think "uncertainty" specifically in terms of Derrida's important theorization of "undecidability." Sullivan's argument—there is no absolute certainty upon which we can base our decisions—shares a dangerous affinity to the traditional poststructuralist argument that we live in a decentered universe, inhabiting a world that can never offer the ontological certainty of an absolute ground. Or, to speak like Derrida, the lack of an absolute order of stability presents us with an irreducible undecidability. This philosophical concept of undecidability is often grossly misunderstood to be a form of liberal relativism, as if Derrida were arguing that in the final analysis, we cannot decide. Nothing could be further from the mark. For Derrida, undecidability is a name for the moment when a subject—in the act of deciding—unchains itself from constrictions of an existing program of already determined relations: "No responsibility is taken if at a given moment one could not decide *without knowing*, without knowledge, theoretical reflection, the *determinate* inquiry having encountered its limit or its suspension, its interruption. Without this interruption . . . there would never be a decision or responsibility, but only the deployment consequent to a *determinate* knowledge, the imperturbable application of rules, of rules of known or knowable, the deployment of a program with full knowledge of the facts."[43] If there were 100 percent certainty, there would be nothing to decide. The field of existing relations (100 percent known) would then be the dynamic force making the decision for the subject. The subject would only have to follow the plan of action given over by the ruling situation. If a genuine decision is to occur, if a subject is to decide without everything already having been decided for him or her, then the subject must decide without, to a certain degree, knowing; he or she must make a choice without full knowledge of the situation. Otherwise, the subject gives over the power of deciding to a set of knowns that control the totality of choices, and no real decision is ever made by the subject.

Derrida is equally emphatic that this relation to nonknowledge does not mean that the subject should relinquish knowing in favor of something like faith or ignorance: "For there to be decision and responsibility, I am not saying that one needs ignorance or some form of not-knowing; not at all, on the contrary, one needs to know."[44] Put differently, the relation between making a decision and undecidability involves an act of cutting, so to speak: "One needs to know as much as possible, but between one's knowledge and the decision, the chain of consequence must be interrupted. One must, in some way, arrive at a point at which one does not know what to decide for the decision to be made. Thus a certain undecidability . . . is the condition or the opening of a space for an ethical and political decision."[45] In this sense, "undecidability" is not a reference to some form of indecision but names instead a space of nondecision that can, like "the part of no part" as identified by Rancière, also be likened to a space beyond the zone of experience. The word *decision* originates from Latin, meaning literally "to cut off"—*de-* ("off") and *caedere* ("to cut")—and the idea to cut stems from *caedere*'s specific relation to *caementa*, which are "stone chips used for making mortar"; the stone chips come from the act named in *caedere*: "to cut down, chop, beat, hew, fell, slay."[46] Put differently, the decision is the groundless act of cutting oneself off from the cemented zone of experience that governs existence. To sharpen this one step further, undecidability refers to that which has not already been decided, and the leap away from the already decided—the encounter with undecidability that has not been reduced to something decided—makes an authentic, not an imitative, act possible.

The difference, therefore, lies in the relation to necessity. In a state of existence in which everything is governed by an Accidental state, the freedom to choose that arises out of a relation to uncertainty or an unknown is turned around so as to affirm all the more the empiricism of necessity. The military commander, the scientist, and the ecocritic all share a relation to a fundamental "lack of perfection," to quote Sullivan. But for the military commander, this lack of perfection is not a source for a groundless act of decision making; it is incorporated into the Accidental so as to strengthen the power of an instituted program of action. Uncertainty and the lack of perfect knowledge are reterritorialized so as to generate a state of what might be called "paranoiac anxiety" (as opposed to an abyssal Kierkegaardian *angst*), which is then used to justify the institution of a state of necessity that positions itself beyond any form of debate—whether that debate is political, legal, or critical-theoretical. Neither can this position of necessity be understood according to any known form of reasoning, conservative or otherwise. It turns the openness given by the event of uncertainty into an unquestionable and insurmountable absolutism. Again, the line between this Accidental state of necessity (which, it must be kept in mind, should not be equated too easily with classical reason; necessity trumps *any* form of reason in this structure) and the space of uncertainty that offers the potential for uncementing oneself from the ruling order (which also poses fundamental challenges to forms of reasoning) is thin and difficult to trace. But the difference could not be more substantial, and it is this difficult difference that will come to define our ecological future with greater decisive force.

From within the ideological position of the de facto state of necessity, the leap from environmental activism to national security is swift: "While the developed world will be far better equipped to deal with the effects of climate change, some of the poorest regions may be affected most. This gap can potentially provide an avenue for extremist ideologies and create the conditions for terrorism."[47] From this position, the leap from environmental activism to postcolonial nation-state warfare is even faster: "Many governments, even some that look stable today, may be unstable to deal with these new stresses. When governments are ineffective, extremism can gain a foothold." In these lightening moves, the complexity and diversity of the current environmental occasion is reduced to a single concern: national and global security. Issues such as biodiversity, animal rights, sustainability, bioengineering, threatened habitats, and so on no longer appear as part of the arena of human political and existential concerns. These calculative moves, which work by preying on the fears of geopolitical insecurity ("extremism"), are designed to camouflage the violence of a reductive logic that shrinks all environmental concerns to the world of "security policy."

Thus the military demand for action—a mode of action that presents itself as an unconstructed, matter-of-fact empiricism that puts an end to debate—is precisely the line of reasoning we should reject. Its relation to an empiricism based on a state of necessity makes it differ in a secondary fashion from the forms of ecocriticism performed in the humanities and social sciences. Ecocriticism in these disciplines takes the exploration of possible forms of cutting open limiting forms of institutionalized decision making as one of its primary concerns. In doing so, it foregrounds the ideological nature of any ecological concern, as opposed to the military demand for action, which conceals its complicity with ideology in part through the use of narratives of neutrality.

Consider the argument presented by Vice Admiral Richard H. Truly, former NASA administrator and the first commander of the Naval Space Command. (The Naval Space Command was established in 1983 during the Reagan administration and is the navy's institution of global surveillance, using extensive satellite observation to support naval action around the planet. It also includes a "space watch" that operates around the clock to track satellites in orbit; it generates a "fence" of electromagnetic energy that "can detect objects in order around the Earth out to an effective range of 15,000 nautical miles." The Naval Space Command operates "surveillance, navigation, communication, environmental, and information systems" in order to "advocate naval warfighting.")[48] In his articulation of his particular stance on the environment, Truly deploys a well-known form of representative transparency:

> I had spent most of my life in the space and aeronautics world,
> and hadn't really wrestled with [environmental issues]. Over the
> course of [a] few years I started really paying attention to the data.
> When I looked at what energy we had used over the past couple of
> centuries and what was in the atmosphere today, I knew there had
> to be a connection. I wasn't convinced by a person or any interest
> group—it was the data that got me. As I looked at it on my own, I
> couldn't come to any other conclusion. Once I got past that point,
> I was utterly convinced of this connection between the burning
> of fossil fuels and climate change. And I was convinced that if we
> didn't do something about this, we would be in deep trouble.[49]

Truly's argument attempts to ground itself in an empirical verifiability that arises outside of human fabrication and ideological influence. *Data* as an object in existence appears without human generation; it arises from outside of any narrative construction and is magically represented without any representative tools. No human speaks to the military commander about climate change; the data "speaks for itself." This fantastic ex nihilo argument, however, undermines itself in the symptomatic pressure put upon Truly to deny twice the existence of any author or organization that might have a relationship to the collected knowledge of climate change: the admiral "was not convinced by any person or interest group." Nor did the admiral encounter any other human than himself during the process of coming to understand the data. This dynamic of data "speaking for itself" conceals a design meant to trigger specific behaviors and results—key among them is the construction and preservation of an irrefutable state of necessary military action.

This assumption of (a)political direct action enables the security state to engage in quasi-wartime measures during a time of peace, putting the *état de siege* into practice. Natural disasters, for instance, have traditionally been thoroughly interpreted in the traditional sense of the term *accident*. They are now entirely understood to be symptoms of climate change and in this sense have become thoroughly enmeshed in the Accidental apparatus. "Weather" is now capable of being harnessed by the security society to support the growth of military power. Witness the case of the 2005 Indian Ocean tsunami. The catastrophic nature of the tsunami is highlighted in a particularly interesting section from the Center for Naval Analysis (CNA) report on climate change and national security titled "Engagement Opportunities." Here, the "humanitarian" role of the military is highlighted: "The U.S. military helped to deliver relief to the victims of the 2005 Indian Ocean tsunami because it is the only institution capable of rapidly delivering personnel and material anywhere in the world on relatively short notice."[50] These two narratives subsidizing the navy's opportunities for engagement—the military's humanitarian mission and its superiority to any other war machine in the world—mystifies the genuine geopolitical rationale that sutures the abyss of third-world human existence opened up by an "unforgiving environment."[51] Weather becomes an opportunity for the U.S. military to ramp up its presence, thereby regaining its supremacy and control in key geographical regions around the planet. The "engagement opportunity" of the tsunami, for instance, generated the new Department of Defense Directive 3000.05 in 2006, which "provides the mandate to conduct military and civilian stability operations in peacetime as well as conflict to maintain order in states and regions" (40). This "exceptional state" rhetoric is coupled with a self-congratulatory, triumphal tone in the claim that only the American war machine could handle such massive environmental turmoil. Thus the tsunami of 2005 became the pretext for extending military power in the "pacific theater."

As Walden Bello points out, U.S. actions in relation to the 2005 tsunami were anything *but* this touted humanitarian engagement: "The relief operations were not a disinterested, peacetime military mission. One immediate sign was the deliberate U.S. effort to marginalize the United Nations (UN), which was expected by many to coordinate, at least at the formal level, the relief effort. Instead, Washington sought to bypass the UN by setting up a separate assistance 'consortium' with India, Australia, Japan, Canada, and several other governments, with the U.S. military task force's Combined Coordination Center at University of Tapao, Thailand, effectively serving as the axis of the entire relief operation."[52] The stealth orchestration of the relief efforts was, in part, an attempt on the part of the Bush administration to repair the damaged image of the U.S. military machine in the wake of the Iraqi War and the War on Terror, which was particularly unpopular to the Muslim majority in Indonesia and the Southeast Asian region in general. It was also a chance for the U.S. Pacific Command—the oldest of the U.S. military commands—to reenergize its waning power and influence in the "Pacific theater," which had diminished after the Vietnam War. The ecological event of the tsunami became a singular opportunity that enabled the military machine to reassert U.S. nationality by taking immediate and decisive action, cutting through any debate or negotiation with the complexities of an international UN peacekeeping coalition. Ecological chaos and devastation became the grounds for quietly reinstalling the former Pacific theater *état de siege*: "a massive expedition that eventually came to encompass over twenty-four U.S. warships, over one hundred aircraft, and some sixteen thousand military personnel—the largest U.S. military concentration in Asia since the end of the Vietnam War."[53] Bello points out that this concentration was cemented further with the reintroduction of military-to-military relations between the United States and Indonesia, a country that had been under a ban of arms sales from the United States and had restrictions on military training due to the effective campaigns of human rights groups that criticized the human rights violations of the Indonesian Army in the 1990s. Less than two months after the tsunami disaster, the military ban on Indonesia was lifted. Congress voted to maintain the ban, but the State Department, "exploiting a national security waiver provision,"[54] overturned this vote so that the military-to-military ties could be strengthened for purposes of "antiterrorism efforts, maritime security," and the ecologically grounded necessity for "disaster relief."[55]

Ecological catastrophe becomes the basis in this militarized environmentalism for reterritorializing the earth as essentially an abyssal milieu—an "unforgiving environment," to use the language of the CNA's national security climate change report—incapable of supporting human existence without the aid of the U.S. war machine. *In this maneuver, the war machine takes over the former role of the ecosystem as supporter and sustainer of human life.* Such a representing of the earth occludes any memory of the enclosure movement and the West's colonial politico-economic relations to nature that have ruled the human–nature nexus from the seventeenth century to the present. The history of the U.S. military presence in the Pacific is steeped in American imperialism and colonial expansion. The U.S. Pacific military structure is the world's largest naval command, with close to three hundred bases and facilities, along with active personnel from the Marine Corps, the Air Force, and the Army, in addition to the Navy. To quote Bello again, "Perhaps the best way to comprehend the U.S. presence in the Pacific is to describe it as a transnational garrison state that spans seven sovereign states and the vast expanse of Micronesia."[56] Typical historical periodizations of U.S. colonial expansion tend to focus on the "errand in the wilderness" activities that engulfed the lives of Native Americans and Mexicans. But the entrance of the United States in the Asian "theater"—with the successful wresting of Guam and the Philippines from Spain in 1898—was also a key part of U.S. colonial history, extending U.S. sovereignty to a global register. The new bolstering of military presence stems from a different type of ideological imperative, an ecoimperial ontology that grounds its rationale in the troubling anarchic essence of nature, which it transforms into an exploitative logical economy that matches the never-ending and centerless essence of "terrorism" as the previous military transcendental signified.

America's prosperity and technological advancement was made possible by inflicting catastrophes elsewhere, outside its borders, in the form of low-wage factories, the support of dictatorships, internecine warfare, the unequal distribution of resources, uneven development, and a mystification process that kept the American soil metaphysically separated from the rest of the planet in the mythology of American exceptionalism.

Throughout all this, it turned a blind eye to the effects of this destructive development in and upon ecosystems across the planet. Its sense of a different historical development from all others conceals its imperial genealogy and ontology. This exceptionalism involves the constitution of a national biopolitical settlement and an autonomous consciousness that lives and sustains its existence far from any connection to nature. If nature exists at all in this paradigm, it comes to presence as the enemy-other that poses a threat to economic development and the security of all people. Today—in the form of global warming—that decision to annex nature to profit machines and colonial apparatuses of power has returned to haunt the imperial logic of exceptionalism.

From Agamben's State of Exception to "Sovereignless" Environmentality: The "Post-American" Security Society

The move to environmentality involves a shift of the institutionalization of a state of exception by the sovereign to the condition of existence itself. In this sense, it is not the nation-state or a sovereign identified with any particular nation-state but a relation of entities more difficult to identify that act (and not necessarily in unison or even clearly orchestrated) by putting into motion a cohesion of concerns that share a certain directed course of action. In the age of climate change, in which nature itself is the ultimate threat across different geopolitical regions, it is no longer the sovereign that suspends the law and extends executive powers; it is a coalition of forces that extend the Accidental, that extend a form of power suspending forms of rationality, the law, democracy, and various possibilities for new ecological relations. Instead of the sovereign suspending the law, it is the military industrial complex of the security society that spearheads the move toward a state of exception. This shift of "full powers" from the executive branch of the nation-state to the more complex confederacy of the Accidental state (military complexes, corporate allegiances, university-industrial centers, private interests, etc.) reveals, in a sense, what was hidden all along behind the fantasy of executive power. Speaking like Foucault, this shift "cuts off the head of the king," who presumably stood behind and controlled the workings of power. More specifically, it makes visible the structure of the Accidental that was the larger ontological structure lurking behind the more "ontically visible" state of exception.

Such a shift away from executive sovereignty to a more generalized state of necessary exception can be seen informing statements such
as those made by the military leaders supporting environmentality. In
Rymn J. Parsons's "Taking Up the Security Challenge of Climate Change"
(a Department of the Army 2009 position statement), he makes this shift
explicit and even provides a military transliteration of the postcolonial
genealogy of empire. In a section titled "The Future Security Environment," he lays out the Accidental narrative of a "post-American" state:

A time is coming, measured in decades, not centuries, in which
American military superpower status may remain, but its relative
economic power may be less and its resulting political prerogatives
may be fewer, that is, a "post-American" world of increasing multipolarity. For nearly 300 years, world order has been shaped by the
hegemony of Western liberalism, first in the form of *Pax Britannica*, and then *Pax Americana*. But a post-American world, though
globalized, may also be more non-Western. It will be a world of
evolving modernity to which the United States must adapt, not
a world the United States will dictate. It will be a world in which
China, already the second-most-important country in nearly
every respect, will take a decidedly American tack, though not by
employing American methods, to expand its influence in hopes
of molding the international system to suit its interests. Further,
it will be a world in which a healthy international community will
still be a vital U.S. interest.

The most important bilateral relationships China and the
United States have today are with each other. Strenuous efforts
must be made to keep the U.S.–China relationship nonconfron-
tational and to encourage China to broaden its responsibility for
promoting and maintaining peace and stability. American grand
strategy and military strategy must include ways and means to
achieve these ends. Because the effects of global warming-related
climate change are of as great, if not greater, concern to China
than the United States, the two countries will find much common ground in this arena. China, like America, will perceive the
security implications of climate change, but it remains to be seen
whether it will play a constructive or discomfiting role. The United
States must focus intently on this issue.[57]

In Parsons's exceptionalist, somewhat anxious and quasi-apocalyptic narrative, we begin to see the full force and meaning of the signifier *adapt*, which, before this, appeared to be the military empirical necessity of succumbing to the inevitability of climate change on the grounds of protecting national security. Here, however, *adapt* names the end of American national supremacy in favor of a "multipolarity" of powers—an indication of the expansion of the structure of the Accidental beyond the "control" of a single nation-state. This suggests an extension of "full powers" from the central node of the sovereign to the complex network of the Accidental itself. The futural characterization of the post-American world equally reveals that this discursive military constitution of an ecological imperative is not really about the environment at all but about the effect that climate change will have on the future of global geopolitics.[58] In February 2009, the CNA and the Institute for National Strategic Studies (INSS) produced a panel discussion report on China's sixth defense white paper, issued a month earlier. Though this report is not available to the public, Parson indicates that the CNA and the INSS are concerned about the waning of U.S. supremacy in the wake of China's declaration that it will be an "active and constructive" "key player" in the post–climate change world.[59] Environmental policy therefore becomes the vehicle for constituting new military flows of power in key locations around the globe.

As we have seen, military and security experts give weight to their position on the democratic playing field by identifying as one of their key strengths the ability to make decisions without having to waste precious time waiting for the bureaucracy to produce a conclusive report. In this sense, they set themselves apart from the old nation-state, which lags behind the Accidental state that the military helps bring into existence: "While political leaders on Capitol Hill seek definitive answers about how quickly the world's climate will change, military and national security experts say they're used to making decisions with limited information."[60] This waning of the nation-state within the larger Accidental state should not, however, be understood according to the analysis proffered by Michael Hardt and Antonio Negri. It is worth pausing for a moment in my analysis of neocolonial nation posturing to consider their important analysis of our current geopolitical setting—especially as it relates to the weight they give to the human activity of warfare. In their works *Empire* and *Multitude*, they argue that the form of the nation as we know it is disappearing in favor of more transnational structures of corporate capital formations. Conversely, the "multitude" is also growing—that is, the "living alternative" of resistant peoples that are gaining momentum in reaction to this "Empire."[61] In *Multitude* Hardt and Negri define the geopolitics of the early twenty-first century in terms of a generalized "civil war" that exceeds the traditional parameters of war as an activity waged by sovereigns of nations: "Whereas war, as conceived traditionally by international law, is armed conflict between sovereign political entities, civil war is armed conflict between sovereign and/or nonsovereign combatants *within a single sovereign territory*."[62] War, they argue, which was formerly a state of exception that stood outside of normal political procedures in civil society, has become the norm of global human production. We are in the age of a "global civil war," a concept of war that extends beyond any particular nationality. Far from an exception to the political structure of civil society, war has become "the primary organizing principle of society."[63] Like the military aggression that defines the structure of the Accidental, the generalization of war beyond the former parameters of nation versus nation expands the spatial boundaries of war to a more immaterial concept of an "abstract and unlimited" enemy: the deterritorializing effect of the U.S. War on Terror, for instance, and the terrorist who potentially extends across all foreign territories, even the homeland itself.[64]

Their point is that the present should be characterized in terms of a war that deterritorializes all boundaries of national and social existence. However, this insightful characterization somewhat misses the mark. As we have seen in the rhetoric of the war documents examined in the previous sections of this chapter and chapter 3, the concern of the military is no longer war as we understand that concept, even deterritorialized war. "War" as the primary motivator for military activity is being replaced by something more insidious and ubiquitous: "security." "Security," unlike war—which needs either congressional or executive approval and requires substantive political prestidigitation to win local and international approval—refers to a daily state of existence deemed necessary in order to preserve life. To speak like Foucault, security functions along the lines of a "policing" mechanism; that is, it is central to the totality of procedures that make life and the successful production of life possible.[65] It is understood to be a necessity of existence without which a people cannot live. In this way, "security" can easily be grafted onto the discourse of scientific ecology. Thus the military and the scientific community can readily join forces around the issue of environmental destruction through the combinatory power of the nexus of "security." "War" as an actual occurrence and as an event needing to be planned does not carry this same power to motivate the necessary life and work matters of one's daily existence.

To give an example, recall my earlier discussion of the 2012 National Council for Science and the Environment conference on Science, Policy and the Environment. The theme was "Environment and Security," and, in addition to leading members of the scientific community, the conference highlighted presentations by Sherri Goodman (executive director of the CNA Military Advisory Board), Vice Admiral Lee Gunn (president of the CNA's Institute for Public Research), Rear Admiral David Titley (director of the Navy's Task Force on Climate Change), Major General Muniruzzaman (president of Bangladesh's Institute of Peace and Security Studies), Marcia McNutt (director of the U.S. Geological Survey), Nancy Sutley (chair of the White House Council on Environmental Quality), Rear Admiral Neil Morisetti (commander of the U.K. Maritime Force and the U.K. government's Climate and Energy Security Envoy), and a host of others. Military and other speakers each underscored the urgency of our environmental situation. But each also emphasized that measures needed to be taken—and as soon as possible—to avoid escalations of violence surrounding the struggle for land and resources. Their arguments thus reflected a very different policy than that of the George W. Bush administration and its preemptive declaration of war. Rear Admiral Titley, for instance, repeated in his remarks at the plenary session on "Climate and Disruption" that war was the "last thing we want."[66] Instead what are needed are "security measures"—as was made clear by the titles of a number of panels: "Climate Change and Security: Making the Connections"; "Security in a Changing Arctic"; "Sustainable Security, Fragile States, and Climate Assessment"; "Climate Change and Food Security"; "Impacts of Acidification on Food Security"; "The Water-Energy-Security Nexus"; and "Human Ecology, Human Security," to name just a few.

This intensification of security in these venues is symptomatic, as I have been arguing, of the discourse of environmentality. Unlike Hardt and Negri's conception of "Empire," which is constituted in terms of war, the coming "eco wars" of the twenty-first century will revolve instead around issues of security and the nexus between security and ecological insecurity. In this sense, it is not the machine of Empire governing and administrating human existence; it is the Accidental state. The Accidental state—unlike

GENEALOGIES OF MILITARY ENVIRONMENTALITY II · 229

the corporate Empire, which functions according to the teleological pull of capitalist economic development (neoliberalism)—is orchestrated by a fundamental ecological unpredictability, an unpredictability that extends its domain globally, beyond any national or continental geographic borders. To quote Virilio again, "Once upon a time the local accident was still precisely situated—as in the North Atlantic for the *Titanic*. But the global accident no longer is and its fallout now extends to whole continents, anticipating the *integral accident* that is in danger of havoc wreaked by Progress then extending not only to the whole of geophysical space, but especially to timespans of several centuries, to say nothing of the dimension *sui generis* of a 'cellular Hiroshima.'"[67] The accident of global climate change, like the Cold War nuclear experiments in the Pacific that irradiated the entire planet, encompasses the field of existence not only spatially but temporally as well—not only to the "whole of geophysical space" but also to the "timespans of several centuries" (recall the half-life of carbon dioxide). The ubiquity of an environment in need of defense raises environmental security to the level of an "orange alert" state of existence in a way that war as a human activity does not. In this way, the accident is "integral" to both specific and general forms of existence, crossing species boundaries to affect all forms of life.

Environmentality, the State of Exception, and the "State of Nature"

When Agamben connects the state of exception to the unexamined belief in a prelapsarian state, a "state of nature"—the "kenomatic state" or "emptiness of law" "analogous to the idea of a state of nature"—he taps into a genealogy that we can productively connect to the beginnings of a Western imperial conception of nature, one that stands as the defining liminal point of Western colonization/cultivation and extends all the way to the discourse of security defining the present. The original logic for the decision by Western states to adopt the enclosure movement did not only arise from the new economic demand for private property and "high yields" but also stemmed from a growing sense that culture/cultivation could only truly exist if it went to war against the environment—that is, if it successfully constituted "the land" as an essentially unstable force in need of "improvement" and defensive forms of domestication (meticulously surveyed and measured spaces protected and enframed by stone walls, fences, ditches, hedges, etc.). This second belief brought about a constitution of nature as an unstable, unreliable force—and the colonial territories in the far-flung periphery of the imperial metropolis were subsequently reterritorialized to become the most chaotic and unreliable geographical spaces. The enclosure of the land that rewards the sovereign landlord with increased economic profits in the early international market cannot be thought of apart from this extensive colonial apparatus.

This grounding is a pure state of nature that, by the very character of its supposedly being the starting point of all politics, suggests that the establishment of a state of exception is not an arbitrary move on the part of a power-hungry politician but the return to a more supposedly genuine form of the political. This move matches point for point the movement that we call the logic of scientific empiricism and, as we saw when examining Manuel De Landa's work on the rise of twentieth-century technology, the movement of a supposedly pure mathematical logistics (when the natural sciences and the formal sciences are woven together to work in tandem). The installment of a state of exceptional politics is represented as the ultimate law and an absolute, genuine form of originary politics because it directly springs from the original (pre)law of nature. This is the basis of the war machine's rationale for moving human existence toward adaptation rather than a different set of human–nature relations that would arise from a different "original" form of nature (sustainability or inhabitancy, for instance). The environment is essentially "instability" in the raw, and our adaption to that inevitable instability needs no judicial process to justify itself. Nor does it need any political process beyond the politics of the "fullness of power" attributed to an executive branch now generalized beyond the "fullness of power" of the sovereign to the fullness of natural security, which becomes the systemic *state* of the political itself—the point at which politics as the primary form of human organization shifts to the security society and becomes environmentality. That is, it is the moment when "pure action" arrives, when humans return to the original point of living in a natural, not a political, state.

From Environmentality to an Ecology of Inhabitancy

Territory and possessions are at stake, geography and power. Everything about human history is rooted in the earth.

—Edward W. Said, *Culture and Imperialism*

These white men looked on native life as a mere play of shadows. A play of shadows the dominant race could walk through unaffected and disregarded in the pursuit of its incomprehensible aims and needs.

—Joseph Conrad, *Victory*

By way of a conclusion, I would like to return to Chakrabarty's important point that we need, in the anthropocene age, to expand our images of the human—that is, to think contradictory images of the human simultaneously "on multiple scales and registers" in order to open a sense of understanding that would critically challenge current socioenvironmental paradigms. Alongside the images of the human that Chakrabarty invokes in his essay (the decentered, differential subject and the traditional Cartesian, Enlightenment individual—both of which are necessary to the continual functioning of the neoliberal paradigm), I have suggested that the anthropocene era calls for two additional images: the neoimperial human of modern technology that acts as a geological, securitizing force and the preenclosure inhabitant. The list of postcolonial critiques of the image of the Enlightenment human is extensive and does not need to be rehearsed again here. For this reason, I have attempted throughout this book to resituate the Enlightenment image of the human in a different genealogy so as to reveal the precise manner in which it has become a "geological force" arising in tandem with a security society founded in modern military and technological history. As an alternative and hopefully less ecologically threatening possibility, I have offered the image of the inhabitant—not as an empirically verifiable truth or a utopian solution, but as a kind of strategic rejoinder. As we saw in chapter 2, "inhabitancy,"

231

once highly active as a human identifier before the totalizing shift to common law, serves as an important marker of what capitalism and imperialism had to destroy in order to complete their paradigms.

In this concluding section, I reconsider these sociopolitical developments more theoretically by situating them alongside Agamben's recent meditations on human subjectivity and the environment. Agamben analyzes what he refers to as "the animality of man"—his phrase for an end-of-history, alleged organic humanity first articulated by Kojève and then later embraced by highly influential proneoliberalists like Francis Fukuyama. This critical nomenclature is of particular importance because of the connections we can draw between a posthistorical politics and the mythology of a pure, nonideological state of existence that makes axiomatic militarized deployments of natural and formal modes of intelligibility (organic military thought, the Prisoner's Dilemma, "natural security," etc.). The "animality of man," I argue, is the formal and natural state of biological existence summoned into existence by the militarized milieu of scientific exploration in the age of environmentality. It is in and through this ontological transformation of the human species that the security society, in preparation for the coming resource wars, hopes to achieve an undisputable rule for securing the territory of the earth. The "animality of man" thus serves as an apt descriptor for the biopolitical battle of life that has besieged human existence in the anthropocene—a war of "adaptation" and "natural security" waged against all other human forms of intellectual and aesthetic possibility. In opposition, Agamben retrieves the ecological concept of the *Umwelt*—the radically different conception of the environment first proposed by the biosemiologist Jakob von Uexküll. In this conception, I argue, we might find a different, less ecologically destructive possibility for inhabiting the earth.

The Animality of Man and the *Umwelt*

Agamben's analyses of today's suspensions of the law and the political in both *Homo Sacer* (published in English translation in 1998) and *The State of Exception* (2005) are curiously interrupted by a work called *The Open: Between Man and Animal* (2004). In some ways, *The Open* is a continuation of his concern for a humanity caught up in the violence of political and judicial suspension. But unlike the others, *The Open* contains an analysis of human subjectivity that is ecological in nature and offers examples of environmental relations strikingly heterogeneous to those that are common today. The text spends the bulk of its time concentrating on a contemporary phenomenon that Agamben refers to as "the animality of man." His exegesis on this phenomenon reaches a culmination in a lengthy discussion of Heidegger's conceptualizations of the difference between humans and animals and the influence that Jakob von Uexküll had on Heidegger's development of a more "ecosystemic" understanding. The text thus analyzes two phenomena that are important to our exploration of the security society and its erasure of inhabitancy: the institutional establishment of a posthistorical, postpolitical "human animality" and the radically different environmental relation found in Jakob von Uexküll's conception of the *Umwelt*.

Agamben begins *The Open* by exploring how "man" and "animal" have been discursively constituted in a reciprocal relation throughout history. He identifies this reciprocal constitution as an "anthropological machine." The parameters of what makes one human and not an animal, he argues, have shifted continually throughout the long historical running of this machine. What it means to be human was always a question and, even throughout the rise of the sciences in the Enlightenment, is continually being renegotiated (in the natural sciences, for instance, and its shifting classifications). In the twentieth century, however, there are signs that this shifting relation is coming to an end—being replaced by a kind of collapsing of "man" and "animal" in on one another—resulting in a new status of the human that Agamben names "the animality of man."

The phrase "the animality of man" originates in Alexandre Kojève's *Introduction to the Reading of Hegel* and describes the state of human existence when it reaches the "end of history." Kojève's well-known, teleologically oriented interpretation of Hegel is instrumental. As I argued in the previous chapter, the security society attempts to manufacture a new form of apolitical "freedom" by constructing the fantasy of an organic form of human existence that is supposedly aligned with the ontological essence of nature. Such a conception owes much to the contemporary end-of-history discourse of neoliberalism. Agamben's return to Kojève thus also constitutes an engagement with the philosophical basis of the neoliberal paradigm (which received its philosophical justification in Francis Fukuyama's highly influential *The End of History and the Last Man* in Kojève's reading of Hegel). The following is the passage from Kojève in which this idea of a "post-historical" humanity appears:

If Man becomes an animal again, his arts, his loves, and his play must also become purely "natural" again. Hence it would have to be admitted that after the end of History, men would construct their edifices and works of art as birds build their nests and spiders spin their webs, would perform musical concerts after the fashion of frogs and cicadas, would play as young animals play, and would indulge in love like adult beasts. But one cannot say that all this "makes man *happy.*" One would have to say that post-historical animals of the species *Homo sapiens* (which will live amidst abundance and complete security) will be *content* as a result of their artistic, erotic, and playful behavior, inasmuch as, by definition, they will be contented with it.[1]

This relationship between the Animal and the Human manifests itself differently throughout history. The "anthropological machine" constantly divides "Man" from "Animal" in order to endow both with meaning. The precise manner in which each age draws this division indicates its organizational rationale. However, the nature of this divisionary machine is not to establish an inherent ontological difference between the two. As Agamben argues, the division is never finalized. Moreover, the function of the anthropological machine is to keep the Man–Animal boundary open ended to constitute the division not as a finality but as a space of dispute.

However, for the postpolitical, posthistorical Man ushered in by Kojève, the division has outlived its usefulness. The end of history signals the end of the anthropological machine's necessary performance and the dawn of a state of existence that a humanity steeped in historical context had been incapable of realizing. I would like to identify this as the formal and natural state of biological existence that we have been disclosing as the central teleological goal of militarized scientific exploration in the age of the security society. As is well known, Fukuyama interprets this natural form of existence economically—as the logic of late capitalism.[2] Agamben, however, isolates a destructive gesture that necessarily transpires in the completion of this particular end, one that Fukuyama overlooks: the realization of the posthistorical man "must also entail the disappearance of human language, and its substitution by mimetic or sonic signals comparable to the language of bees. But in that case, Kojève argues, not only would philosophy—that is, the love of wisdom—disappear, but so would the very possibility of any wisdom as such."[3] The end of history that brings about the naturalization of Man *also* brings an end to wisdom—which we can clarify here as the end to the critical questioning of patterns of thought found in the activity of Gadamerian "understanding." Of additional interest here is Kojève's description of the postlanguage human, which, at first, may sound absurd. However, when we set this discussion of the "language of bees" next to the time and energy that the military put into the search for ways to find a natural form of communication through which they could entrain not only soldiers but citizens to act out the esprit de corps of national desire, we suddenly discover that Kojève's seemingly innocuous footnote points directly to the heart of the organic analytic of environmentality that we've been interrogating. It characterizes the precise form of life that the security society hopes to achieve in the twenty-first-century era of diminishing resources—a synthesis of the disciplines of the formal and natural sciences into an undisputable rule for securing the territory of the earth.

Agamben emphasizes the astonishing brutality of Kojève's antiphilosophical philosophy of posthistoricism, which reduces the massive conflicts of the twentieth century to the "alignment" with the naturalized form of the final frontier of existence: "Everything that followed—including two world wars, Nazism, and the sovietization of Russia—represented nothing but a process of accelerated alignment of the rest of the world with the position of the most advanced European countries."[4] Kojève's trips to the United States convinced him that America had succeeded and become "fully aligned" with posthistorical life. America, he argued, was ahead of the rest of Europe, with the Russians and the Chinese rapidly attempting to catch up. American Man was thus naturalized (animal) Man: "The 'American way of life' was the type of life proper to the posthistorical period, the current present of the United States in the World prefiguring the future 'eternal present' of all humanity. Thus, man's return to animality appeared no longer as a possibility that was yet to come, but as a certainty that was already present."[5]

This "return" to animality, however, conceals its ideological underpinnings and inherent violence. Its process of self-justification *introduces an entirely new conception of animality*, one especially designed to meet the demands of the posthistorical struggle for life and resources. Agamben concludes this opening chapter by associating this concept of animality with Foucault's biopower—that is, the identification of human existence in the age of security. It is biopower's single obsession with the battle to produce life that takes over the totality of existence; this is the prime concern of a posthistorical, postpolitical humanity, annihilating all other human forms of intellectual engagement and cultural activity. "Organic" or "entrained" life, "natural security," "adaptation"—these are signifiers defining a SAGE type of intelligibility seeking to synthesize—to entrain—a militarized form of Man and technology with Nature. The "animal" in the human is life purified of (human) ideological, political, philosophical struggle—that is, life supposedly in its most originary form: *animal* life, life understood tautologically as the basic struggle for life. This idea of animality is a given, nonconstructed element of existence that Kojève claims to be a concern of human existence throughout various historical stages of its development. For Kojève, the "human" is an entity *caught within* historical struggle and consequently at odds with the Animal "within" Man. Hence Agamben can declare that "Man" and "Animal" have been—in their political and philosophical forms and so on—not separate entities but a *division*, one "ceaselessly articulated and divided" throughout history.[6] This ceaseless articulation and shifting division ends once "Americanized humanity" reaches the posthistorical age. Possibly the various Enlightenment disciplinary divisions of the human from the nonhuman (biological classifications, natural science's divisions of species, taxonomic tables, etc.) owe their common passionate classificatory procedures to this fundamental search for the answer to what is human and what is animal—in other words, to putting an end to the search so as to insert the human back into its originary posthistorical animality. It is possible too that this also affects theological articulations of the division that have been so important to the founding of religious orders—the body and the soul—in terms of this posthistorical ontology.[7]

The designation of the nature of the human in the current historical occasion therefore involves an imperial claim to dismiss what previous ages have found to be essential about human intelligibility: the capacity for thinking and articulating the anthropological *division*. The effort was not to discover an ontological essence but about the division itself, how and why it was constituted, where it was placed, and for what reasons. The discourse of environmentality orchestrates this inquiry to pin the question down, to empirically verify the presence of an ontological guarantee. The generalized collapse of the human/animal boundary throughout the twentieth century indicates the size and extent of the forces at work attempting to erase the division in the face of planet-wide ecological catastrophe. The worldwide eradication of this boundary is indicative of the manner in which different states (and their contradictory ideological outlooks) will seek to "cooperate" with one another. Agamben's book suggests that there are different methods by which we can bring back into play the question of the human outside the metaphysics of a postpolitical theology. He suggests that we think of Man in more ecological terms: outside the terms of the kenomatic, biopolitical life as authorized by environmentality. After his initial discussion of Kojève, the bulk of the remaining text concentrates on Heidegger's enactment of the anthropological machine. Heidegger's access to the question—and by extension to human–ecological relations—is made possible, Agamben argues, by his reading of the Baltic German biosemiotician Jakob von Uexküll. Despite the influence Uexküll had on Heidegger (and later Deleuze), his work is barely discussed today. What fascinates Agamben is Uexküll's conception of the *Umwelt*—one of two words for "environment" in German. *Umwelt*, however, is a complex and rich term and is suggestive of much more than its English equivalent. One cannot emphasize enough Agamben's analysis of this term, which is on a par with Freud's thematization of *unheimlich* in the foundational essay "The Uncanny."

Umwelt refers to the limited and defining spatiotemporal relations peculiar to a living creature and its coordinates: "Too often . . . we imagine that the relations a certain animal subject has to the things in its environment take place in the same space and in the same time as those which bind us to the objects in our human world. This illusion rests on the belief in a single world in which all living beings are situated."[8] For instance, space and time do not exist in the same manner for a human as they do for a horse or a flea. The *Umwelt* of each is fundamentally different and carry different significances. Nor can we say even that the *Umwelt* of a race horse is the same as that of a horse employed in the pulling of a cart. In fact, according to Uexküll's logic, the essence of a horse involved in the act of working (not racing) would actually be closer to the essence of an ox in the significances that constitute its environment.[9] Hence even two animals of the same species can be worlds apart from one another when it comes to their concrete relation to an environment.

This last example is actually an observation of Uexküll's highlighted by Deleuze and Guattari in their work *A Thousand Plateaus*. It's worth mentioning their reference here because of the strategic manner in which they frame their allusion to Uexküll in the context of a discussion about ethics.

The allusion is made specifically to serve as an example of their "affective" methodology, an action-conception that endeavors to escape from the limitations of Enlightenment practices of classification and capitalist labor relations. It is worth quoting the passage at length:

> We will avoid defining [a body] by Species or Genus character-istics; instead we will seek to count its affects. This kind of study is called ethology, and this is the sense in which Spinoza wrote a true Ethics. . . . Von Uexküll, in defining animal worlds, looks for the active and passive affects of which the animal is capable in the individuated assemblage of which it is a part. For example, the Tick, attracted by the light, hoists itself up to the tip of a branch; it is sensitive to the smell of mammals, and lets itself fall when one passes beneath the branch; it digs into its skin, at the least hairy place it can find. Just three affects. . . . It will be said that the tick's three affects assume generic and specific characteristics, organs and functions, legs and snout. This is true from the standpoint of physiology, but not from the standpoint of Ethics . . . [w]e know nothing about a body until we know what it can do, in other words, what its affects are, how they can or cannot enter into composi-tion with other affects, with the affects of another body, either to destroy that body or to be destroyed by it, either to exchange actions and passions with it or to join with it in composing a more powerful body.[10]

For Deleuze and Guattari, the importance of Uexküll (and their analysis of relations in general) lies in the thinker's concern for the *compositional* nature of living entities and their environment. I use the term *composi-tion* here in its musical sense: a combination of heterogeneous notes and rhythms. It is a use that I think Deleuze and Guattari would have approved, for they later characterize Uexküll's concept of the *Umwelt* in terms of musical counterpoint—as the combination of nature with music. The question of composition is the question of "what a body can do"—a body's abilities and limitations to compose itself in and through relations with an exterior world, its capacity to be destructive of other bodies or other bodies destructive of it, or the possibility of joining with another body to become a different kind of "machine." These are affective move-ments beyond any kind of facile classificatory schematics. It is a question instead of establishing nondestructive relations to the entire *heterogeneous otherness* of different worlds and their affects. Or, to use an example from Salman Rushdie's articulation of this absolute otherness in his novel *The Ground beneath Her Feet*, "We should try to experience reality as a bat might. The purpose of the exercise being to explore the idea of otherness, of a radical alienness with which we can have no true contact, let alone rapport. . . . Bats live in the same place and time as we but their world is utterly unlike ours. . . . And there are many such [worlds], believe me. All these bats, all of us, flapping around one another's heads."[11]

I have identified this conception of a singular set of relations before—where "every environment is a closed unity in itself"[12]—in the term *local absolute* and in the idea of different "gaits" measured by Reynolds numbers. We can also resituate this discussion of the *Umwelt* in terms of its effect on a generalized order of representation: the shift, for instance, between the singular heterogeneity of custom to the nationalized (and internationalized) common law. The universal nature of common law attempts to impose an objective set of spatiotemporal relations on different environments. The singular heterogeneity, or local absolute nature of the concept of the *Umwelt*, introduces the salient feature that there is no universalized "environment" per se. Rather there are only a series of environments or, more precisely, environments that are not part of a series but unities unto themselves. The generalized human—that is, the "animalized" posthistorical human that acts as a geological force—imposes a higher-order machine onto a heterogeneity in order that it may subsume its varying relations within a command structure, within a larger system of significance (C3i), which is in turn given the imprimatur of the (militarized) natural and formal sciences. The command, control, communication, and intelligence representative apparatus attempts to establish a calculative total system, or total world view. Uexküll identified this totalizing view of the world with the signifier *Umgebung*—the second German word for "environment," which means "the objective world." But by "objective" Uexküll did not mean "organic" or "true." He identified "objective" in terms of *human* understanding. The "objectivity" of the *Umgebung* has significance only for humans and as such is *their Umwelt*. We thus see in Uexküll's identification of the human "environment" as *Umgebung* rather than *Umwelt* the awareness of the human as a geological force: the human (at least the twentieth-century human—Uexküll is, after all, writing about humanity as it exists in the industrial age) is defined as the animal that disrupts the singular heterogeneity of the planet's *Umwelts*. I find in Uexküll's conception of these two very different ecological conceptions a productive theory of the contradictory "multiple scales and registers" of human subjectivity that Chakrabarty defines as *the* intellectual challenge of our occasion.

In terms of the task of this book—the influence of the security society on ecological relations and conceptions—the many different uses of the two designators are striking and tell us something essential about their oppositional nature. Both *Umgebung* and *Umwelt* are translated into English as "environment," but their uses could not be more extraordinarily suggestive: the former of the militarization and geological homogenization characteristic of posthistorical humanity, the latter of preenclosure inhabitancy. *Umgebung*, for example, appears in such phrases as "familiar surroundings" (*gewohnte Umgebung*), "hostile environment" (*aggressive Umgebung*), "stomping ground" (*alte Umgebung*), "hazardous environment" (*gefährliche Umgebung*), "nonconducting environment" (*nichtleitende Umgebung*), "test environment" (*Test-Umgebung*), "on one's own ground" (*in vertrauter Umgebung*), "to be acquainted with the environment" (*die Umgebung kennen*), "fenced off from the surroundings" (*mit einem Zaun von der Umgebung abgeschirmt*), "IT (information technology) environment" (*IT-Umgebung*), and "IEC System for Certification to Standards Relating to Equipment for Use in Explosive Atmospheres" (*IEC-System zur Zertifizierung nach Normen für Geräate zur Verwendung in explosionsgefärhdeter Umgebung*). These uses clearly resonate with the effects of ownership (familiar, stomping), knowledge about (familiarity, acquainted), enclosure (fenced off), the threat to security (hazardous, hostile), and a scientific use (nonconducting, information technology)—all military and targeting concerns.

Umwelt is found operating elsewhere, in another, more singular and nonmetaphysical domain: "negative environmental impacts" (*nachteillige Auswirkungen auf dei Umwelt*), "to be closer to the ground" (*in direkterem Kontact mir der Umwelt stehen*), "Humans are informed by their environment" (*Menschen warden durch ihre Umwelt geprägt*), "concerned about the environment" (*in Sorge um dei Umwelt*), "may cause long-term adverse effects in the environment" (*Kann längerfristig schädliche Wirkungen auf die Umwelt haben*), "Environment protection first!" (*Schützen Sie die Umwelt!*), "marine environment" (*marine Umwelt*), "specific environment" (*nahe Umwelt*), and "nature and nurture" (*Anlage und Umwelt*).

For Uexküll, the anthropological *Umgebung* is ultimately a fantasy, even though it has real effects. In the anthropocene age, the human imposition of an *Umgebung* has become a will-to-power over the earth. Put differently, the phantasmatic structure of the *Umbegung* is a will to eradicate an awareness and engagement with domains from the standpoint of their own inviolable occasion. It constitutes a will to eradicate inhabitancy. To quote Agamben again, "There does not exist a forest as an objectively fixed environment: there exists a forest-for-the-park-ranger, a forest-for-the-hunter, a forest-for-the-botanist, a forest-for-the-wayfarer, a forest-for-the-nature-lover, a forest-for-the-carpenter, and finally, a fable forest in which Little Red Riding Hood loses her way."[13] In opposition to the posthistorical animalized human that acts as a geological force, we might find a more sustainably instructive method of existence through practices based on relations to environments as *Umwelts*—namely, by acting as beings aware of the fundamentally different set of relations that obtain in the act of inhabiting. Such a view more closely matches Barry Commoner's advancement of the classical, ecosystemic view of biology over the target-oriented methodology of molecular biology.

If this important intellectual activity of "contradictory simultaneity" is to be genuine and not allow one of these designators to uncritically subsume the other, it must also open both terms to questioning. Agamben, for instance, complicates Uexküll's meditation on the *Umwelt–Umgebung* difference by considering Heidegger's theorization of their ontological difference. Heidegger's conceptualization of the human–animal division stems from a dismissal of the classical metaphysical constitution of the *animal rationale*. His thematization of the animal complicates any attempt to uncritically identify the *Umwelt* as the single solution for creating sustainable ecological relations. Despite the power of the concept to challenge the totalizing nature of military higher-order machines, its very emphasis on the local poses serious limitations: its attachment to localized relations— that is, its absolute otherness—does not allow for the recognition of other *Umwelts*. It is, in other words, a conception of otherness that cannot itself recognize otherness. In its "passionate attachment" to its own otherness, it is foreclosed from anything outside itself. What Uexküll defines in terms of a set of relations only having "local significance," Heidegger defines more critically in the syntagma *disinhibiting ring*.[14] As a psychological term, *disinhibitor* names a lack of restraint: "The animal is closed in the circle of its disinhibitors just as, according to Uexküll, it is closed in the few elements that define its perceptual world."[15] For Heidegger, "the animal" is the being associated with an *Umwelt* environment; it is a form of being that cannot unchain itself from its passionate attachment to a limited and limiting set of relations. "Captivation" is the "animal's" peculiar state of being or condition of possibility; its "intelligibility" is constrained by this specific environment.

This encompassing mode of captivation perfectly describes the character of the human in the anthropocene. In the anthropocene age, human intelligibility is defined by its inability to generate and maintain an awareness of different worlds. The "animality of man," in other words, signifies a collapse of world making (the human *Umgebung*) into a monstrous form of an *Umwelt*—the human no longer capable of unchaining himself or herself from the specific conception of his or her worlded environment. The posthistorical "animalized human" of environmentality is the being that generates and securitizes a single *Umwelt* that has burst forth to become a planetary geological force. Resituating this in the terms we have been using to address the structural question of transcendence, we can add to Heidegger's definition of "the animal" here (or what I am calling, after Agamben, the "animalized human"): The animal is that entity that has no relation to the sublime. Its ability to transcend remains idle because its flow of life is limited to the securitized relations contained within its disinhibiting ring. Life is only a mode of existence realized as and in captivation—a subjection to a self-contained environment.[16]

Heidegger puts great emphasis on the animal's captivation to its limited *Umwelt*/environment: The animal is subject to a certain structural drivenness from which it has no ability to inhibit its mode of being. It is *taken* by its environment but cannot take a critical position over and against this environment. It is in this sense that Heidegger states that the "animal can only behave insofar as it is captivated in its essence. . . . Captivation is the condition of possibility for the fact that, in accordance with its essence, the animal behaves within an environment but never within a world."[17] More specifically, the character of relation for the animal is one of openness—it opens itself to a connection to its *Umwelt*—but this "openness" takes a particular form, one we might call an excessive ingenuousness, whereby the animal cannot open the relation further to other potentialities. Heidegger further articulates this excessive ingenuousness by stating that the animal's ontological status is one of being open (*offen*) but not openable (*offenbar*). Being able to open one's self to other relations is denied to the animal: the animal exists in an environment but cannot open itself to other environments—that is, it cannot world itself. The animal is thus "poor in world."

The end-of-history conception of human existence—the presumption of living in an organic state of existence that has escaped all ideological, historical, and political influences—therefore shares something essential with the concept of *offen* that chains the animal to its *Umwelt*. However, it is important to recognize that for Heidegger, there is also what we might call an "animality of the animal," for the animal is also that entity incapable of expanding its *Umwelt* onto others. This is an act of colonization that only the animalized *human* undertakes. The problem, therefore, is to conceptualize forms of "worlding"—of creating an *Umgebung*—that do not replace the absolute ecological heterogeneity of the planet's many *Umwelts* and that also do not erase the ability to be open or *openable* (*offenbar*). In this sense, we might attempt to offer a different image of the human, one more ecological or ecocritical in nature and more capable of sheltering and protecting the singularity of the planet's *Umwelts*—to counter the violence of neoliberalism, which attempts to crack open any and all the planet's *Umwelts* to the global flow of capitalist production, imposing *speed* upon a series of heterogeneous *gaits*, no matter what the cost. With such a conception in mind, we might better understand some of Heidegger's somewhat cryptic phrases from essays such as "The Origin of the Work of Art" that theorize a difference between "world" and "earth" and the manner in which he encourages an image of the human capable of "producing the earth" and "saving the earth" by way of "sheltering" and "concealing": "To produce the earth means: to bring it into the open as that which closes itself in itself."[18] The "earth" as a living entity operating essentially in terms of "closing" does not mean that it enforces "the rigid insistence upon some contingent state."[19] Rather, the closing nature of the earth names an ethical activity, one in which the *protection of specificity* is allowed to come into existence and maintain its existence against the violations of a world that would seek to colonize anything and everything. The point in rehashing all this seemingly mystical language is to recognize and not dismiss the peculiar singularities of ecosystems—to acknowledge their distinctiveness and the danger of destroying that distinctiveness for reasons of some higher imperial goal that would want to entrain all for the purposes of realizing a single-minded, neoliberal end. In this sense, posthistorical Man is that creature who attempts to soar away from the earth, who sees the earth and its geographical distinctiveness as a burden to be overcome by technological advancement, speed, dominance, and security.

Is it any wonder that a perverse concept such as "natural security"—the animalistic struggle for existence in and through the struggle for physical resources—would come into existence in the midst of the greatest planetary change experienced by humans? In this sense, "natural security" and "energy security" are two sides of the same coin: they are phrases for the wrestling of resources from every other species on the planet. And life constituted in the discourse of biopower—the improving of human health, the extension of longevity, the strengthening of bodies, the fostering of greater populations, and so on—is the first stage in the process of redirecting human thought toward the postpolitical struggle against the plenitude of nature, of what will go by the name of nature after experienced species extinction. In the age of climate change—which is the result of the predominance of the human appropriation of the planet's stored resources, its various *Umwelts* for its own *Umgebung*—the negative impact of this *accident* of animalized Man reveals its essence in the militarized management of planetary life. It is in this sense that we should understand what Agamben means by "animalization" and his conclusion that, in today's posthistorical occasion, the "total humanization of the animal coincides with a total animalization of man."[20]

To return again to my discussion of Rancière at the end of chapter 1, the environmentality of the animalized human *Umgebung* is the fantasy of a policing movement that is obsessed with accounting for all, with nothing left over. The simultaneous contradictory image of the *Umgebung–Umwelt* (like Freud's discovery that contradictory meanings of both "at home" and "not at home" exist at the same time in the German word *unheimlich*) *names a conception of the universe as made up of habitations that conflict in their manner of counting the world.* The forest-for-the-indigenous-Amazonian-tribe is incompatible with the forest-for-the-neoliberal-timber-company. But the forest-for-the-indigenous-tribe does not attempt to make its *Umwelt* an empire to rule over others, as if it were a transnational corporation. At such a point we may realize that the entire orientation of neoliberal sustainability is opposed to the protective and ethical idea that certain forms of movement and development are fundamentally at odds with the survival of habitations.

The collapse of the human and the animal is the culmination of a species change that began with the erasure of inhabitancy in 1608. The "anthropological machine," despite its uneven and often violent history, has always kept the question of the boundary of the human open. Today that machine no longer functions.[21] As we have seen in the previous two chapters, the human species has moved closer to its captivation by a certain essence of technological development, one that works in tandem with the economy of capitalism, the actions of the military, and the (a)political policing policies of the security society: "Posthistorical man no longer preserves his own animality as undisclosable, but rather seeks to take it on and govern it by means of technology."[22] The ready-made compatibility of these three is a symptom of the foreclosure of inhabitancies. The essence of technology, despite its various justifications throughout the twentieth century (as a necessary means to end the war, as an unavoidable answer to the Prisoner's Dilemma of the Cold War, and as a neoliberal-colonial means to "help" the developing world), has been the mastery of nature, a mastery that has been made possible by the patronizing annexation of the nonhuman *Umwelts* to the zone of nonbeing and nonknowledge. This is the essence of environmentality, of human animality, and of an Accidental architecture operating at full capacity so as to generate the age of climate change.

The Name of Inhabitancy

As I have been suggesting throughout this book and throughout my discussions of Agamben here, an ecological conception of inhabitancy is one productive alternative for reimagining human subjectivity in the age of the anthropocene. If being human means having the potential to transcend the habitation of one's *Umwelt* while simultaneously attending to the radical and incompatible alienness of various habitations, then we might reconceptualize the human as an entity given over to a structure of responsibility where the ethical and the political cross one another—where the creation of a historical destiny (always a function of the human) must come with a politics aware of the ethical consequences of building an environment that will necessarily affect others. Here too history crosses with geography. No nonhuman animal can transcend to such an extent as to make the entire planet its domain. No animal can sit idle and bask in the glory of technological advances that have put an end to the "drudgery of manual labor" only to pursue the accumulation of luxury goods at the expense of others. Yet this is precisely what the animalized human has become. Against this we might posit a conception of the human as always moving through different incompatible habitats and gates of existence—the infinite habitats of forests, grasslands, plateaus, rivers, and oceans, not to mention the habitats of subaltern human labor made invisible by the fetishization of extravagant commodities in the Global North and West. This would also be a conception of the human as an ecological entity thinking the limitations of its various modes of intelligibility in relation to the new afflicted materiality and thinking of the twenty-first-century generation of animals: the thinking of a polar bear without ice, of an albatross in oil, of a bat without night. Such thinking envisions an ontological structure of being "openable" to but not destructive of different environments—a structure made possible by the double play of two contradictory images of the human in which we see positive and destructive elements in both *Umwelt* and *Umgebung* conceptions.

Despite the historical differences between the erasure of inhabitancy in the past and in the present, there are similarities that cannot be denied. The enclosures of the eighteenth century came to be regarded, after centuries of protest, as a revolutionary scientific development. Eventually the naturalization of intense efficiency and economic productivity made the enclosing paradigm appear as if it were an organic state of ecological existence. But this efficiency came at great cost. By the mid-eighteenth century, more than 43 percent of English land was under the ownership and control of fewer than 0.2 percent of the population.[23] Normalized over the course of the next two hundred years, this historical pattern of ownership, domination, and dispossession is now seemingly an irreversible fact of life. According to the United Nations, there are currently 1.6 million inadequately housed people across the world and an estimated 100 million completely homeless.[24] A mere 2.5 percent of landowners control nearly three-quarters of all private land on the planet. In Brazil, 5 percent of the population owns 80 percent of the land, and the poorest 50 percent and the wealthiest 1 percent have roughly the same share of the national income.[25] There the struggle for inhabitancy, as we saw, has now risen to more than a million in membership (and here I am only counting the landless of the MST [Brazilian Landless Workers Movement, or Movimento dos Trabalhadores Rurais Sem Terra]). Enclosing tactics are often reflected in state economic policies as well, which often worsen the situation in metropolitan in addition to rural regions, as in the case of Mumbai when, during the months between November 2004 and February 2005, ninety thousand dwellings were demolished, rendering four hundred thousand homeless and without provision of settlement.[26] This same situation afflicts the Durban Shack Dwellers Movement in South Africa, which deals with dispossession daily. In addition, the struggle for inhabitancy also unevenly impacts communities along gender lines. In Kenya, where less than 5 percent of Kenyans own land and where men traditionally own access to the land, women—who bear the burden of developing and maintaining the home—have no legal right to home and land accessibility. Such issues are global. Contemporary instances of "land grabbing" especially afflict the Global South, but according to the Transnational Institute (TNI), land concentration and land grabbing has been accelerating in recent years throughout Eastern Europe, increasing stress on farmlands and generating more displaced and homeless.[27] Here in America close to a million people are homeless at any given time, with some 2.5 to 3.5 million homeless over the course of a year.

It is increasingly apparent that climate change will exacerbate these conditions. In terms of environmental impacts, the UN high commissioner for refugees António Guterres predicts that climate change will become the "biggest driver of population displacements."[28] In 2008 some thirty-six million were displaced by natural disasters, twenty million of whom were forced to move because of climate change–related factors. The International Organization for Migration (IOM) forecasts that we will see anywhere from twenty-five million to one billion environmental migrants by 2050 (the most widely cited estimate is two hundred million).[29] The Environmental Justice Foundation estimates that some 10 percent of the planet's human population could face displacement from climate change due to altered weather patterns, water stress, and sea level rises.[30] Each year the numbers of climate change refugees grow. In 2010 more than thirty-eight million people had been suddenly displaced by climate-related natural disasters alone. This figure does not include displacements brought about by the more gradual environmental changes occurring daily planet-wide. In a 2013 the Environmental Justice Foundation declared that "climate refugees now outnumber refugees fleeing persecution and violence by more than three to one."[31] To these immense—"geological"— figures, we would need to add the manifold disruptions and displacements of nonhuman habitations that are now becoming the norm for ecosystems around the planet. Can we say any longer that the policies of enclosing and of prioritizing anthropocentric concerns above all others is a legitimate foundation on which to organize human and nonhuman existence?

Such figures stagger the mind and gravely afflict our customary abilities to imagine alternative forms of existence. These massive amounts of displacement, now increasingly well-documented and rightfully recognized as one of the most urgent concerns facing twenty-first-century humanity, are typically considered and addressed as if having nothing to do with the paradigmatic technological shifts of the past. Hence the necessity to connect matters of climate change to the pervasive environmental impact of modernity's enclosures—perhaps *the* most important event of the modern era in need of study and criticism if we are to discover viable alternatives to the transformation of the human species into a geological force. As I argued in chapter 2, it is precisely against this sea change in human–environmental relations that the naming of inhabitancy first occurred in the seventeenth-century courts. The "naming" of "inhabitancy" should consequently be explored as a dynamic arising out of this initial, singular context of the event. To speak like Rancière, the nomenclatural designation of inhabitancy at a signal point in the historical development of Western humanity embodied the "naming of a wrong." That wrong, in a word, was the legalization of dispossession. What is thus valuable about the term is its singular and active dispersal as a counterforce during a time of sweeping social and ecological change. The point in retrieving this concept is not to prove that the inhabitant is a more authentic or a more "natural" form of human subjectivity. The value of retrieving this concept of subjectivity lies in its status as what Alain Badiou would call a "truth-event"—that is, it is symptomatic of the restrictive nature of sociopolitical reality, and it is at the same time a reaction-resistance against that reality. It is an experience that exceeds what is thought to be possible and appropriate, engendering the condition for the possibility of a new way of thinking that moves beyond the restrictive givens of a situation.[32]

Hence the importance of talking about the inhabitant and the dynamic of inhabitancy as a *naming*. In using the preparatory phrase "the name of," I take my cue from Bill Reading's productive syntagma "the Name of Thought"—a phrase designed to counter the triteness and emptiness of the term *excellence*, as widely used today in academic settings. *Excellence*, he argues, is an empty word administrators use for marketing purposes. To this he opposes the term *thought*. Being a good poststructuralist, however, Readings knows that *thought* as a term is also empty of any positive content. *But* its emptiness functions very differently. Like any word, its value lies in its appearance in a context and in the way it takes a position over and against that context. For Readings, the phrase "the Name of Thought" is designed to foreground the contentless nature of thought as a provocation, one that functions "not as an answer but as a question."[33] It is an act of identifying (naming) that breaks open and stops the circulation of naturalized meaning and cultural capital in the existing social order—therefore calling for a conversation about the reigning state of affairs. It parallels the difference that Gadamer marks between *explaining* and *understanding*. In this sense, the naming of inhabitancy in the midst of ecological struggles constitutes a *performance* and should not be mistaken for signifying an ontological essence—much like the performative syntagma of Brazil's largest indigenous organization, the Articulation of Brazilian Indigenous Peoples (ABIP). The rendering-speaking of the word—its "articulation"— functions to call our attention to the regularized state of affairs. The name of *inhabitancy* thus comes to take on a similar but different meaning than it had when the commons system was in full swing. It operates as a resistant designator for the limitations and violations brought about by the reigning circulations of intelligibility and materiality. One could even question if the word was ever that important during the time of the commons system before the advent of the enclosure movement and its massive dispossessions. Most likely it would barely have needed to be uttered. This reveals the inappropriateness of any attempt to return to some hallowed idea of inhabitancy as if it were a positivist, truer form of being human. It is in the act of its naming, in the struggle against the nationwide privatization of the environment, that we should locate the word's special interest and power. Its naming marks the opening of the political: from the moment when dispossessed farmers banded together and began to demand their political right to be inhabitants in a livable habitation. It is in its act of naming a wrong—a speaking by a part that no longer has a part—that constitutes its originative power and its relevancy in the present. The acts of the MST, the ABIP, and other landless and threatened communities—*Via Campesina*, the Rural Women's Movement of South Africa, the Landless People's Movement of South Africa, the Dalit Movement in India, the Peasant Movement of the Philippines, the Hands Off the Land Movement in Europe . . . the list is seemingly and encouragingly endless—are precisely a repetition-return of the truth-event of this name. For inhabitancy, if it can be defined in any succinct manner, is the specter that haunts the policing-politics that have generated the increasingly enclosed, toxic, and militarized world of environmentality.

The Scale of Environmentality

As a planetary event, climate change upends strategies of comprehension and defies explanation by representative orders of interpretation. As I endeavored to show in chapter 1 vis-à-vis the concept of the Accidental, anthropocentric climate change thrusts upon the human species a radically new kind of problem: the inability to transcend a state of existence that promises to confront humans and nonhumans alike for centuries to come. In forwarding a claim on a presumably knowable future, such a statement runs the risk of colluding with the same reactionary apocalyptic enunciations made by the war machine that I have been critiquing. Given, however, the unusual spatiotemporal nature of climate change (worldwide and multigenerational in scale), critical inquiry necessarily risks theoretical contemplation of such a preternatural event while also reflecting on the widespread discourse of predictive catastrophe that surrounds it. While no future is entirely knowable, it is nevertheless unavoidably true that changes in the planet's climate ecology are in the process of unfolding and that these changes will affect the future of the planet for generations to come. The planet-wide and long-lasting character of anthropocentric climate change thus demands that we stretch our customary understanding of the kinds of contextual forces typically evaluated in the humanities and social sciences.

In addition, for instance, to the impacts that global warming will have on the polar ice caps, island nations, food and water security, migration, and weather patterns, the geological force of human-produced ecological changes we are witnessing confront us with the alarming event of an era of mass extinction. According to the latest reports, the planet is in the midst of what ecologists now refer to as the sixth massive planetary extinction of animal and plant species, with as much as 30 to 50 percent of all species dying off potentially by midcentury.[34] This constitutes the worst period of die-offs since the loss of the dinosaurs sixty-five million years ago. To place this specifically within the perspective of human influence: species extinction is a natural phenomenon, with somewhere between one and five species dying each year (known as the "background rate"). Current estimates place the rate of extinction between one thousand and ten thousand times that amount. It is certainly true that the environment will "bounce back" and that new species will arise, but such optimism will have to confront the course of current commitments to mass energy production and consumption, habitat destruction, overfishing, topsoil loss, dam production, pesticides, and a host of other entropic, accidental aftereffects of increasingly geologically consequential human-centered systems of order. Despite the presumptuousness of risking such apocalyptic statements, human-produced mass extinction is an event that cannot be denied and is part of the severe contextual background that critical thought cannot avoid. New solutions will require fundamental changes in these patterns of development.

Greenhouse gases, to push this point further—carbon dioxide, methane, nitrous oxide—are a phenomenon that cannot easily be overcome, at least for generations to come. According to a 2009 study by Susan Solomon, Gian-Kasper Plattner, Rteo Knutti, and Pierre Friedlingstein, carbon dioxide concentration "is largely irreversible for 1,000 years after emissions stop."[35] Visual representations of carbon dioxide and other greenhouse gases have become something of a truism over the course of the last decade. They do not merit repeating for their own sake, but it is nonetheless productive to restage such representative imaginings in order to draw a parallel between the phenomenon of toxic hydrocarbon gases and the measurable planetary influence of military technoscientific development. The following two graphs show how greenhouse gas production is directly connected to the rise and expansion of the industrial age. This firmly links climate change to relatively recent constitutions of humans as geological forces. These constitutions should be set against the many other human formations that can be found in the long history of the species—a history, it should be emphasized, that stretches back several millennia. Inhabitancy is only one of these constitutions—one, in a sense, that can be characterized as being extinct. As the first graph indicates, carbon dioxide levels vary naturally (i.e., without human influence) between 180 ppm and 280 ppm (parts per million). Carbon dioxide and methane have rarely exceeded these natural levels in the course of some 650,000 years. Human beings may only have been a small part of this history, but they have only recently begun to dramatically change it—in roughly the last two hundred to four hundred years—thus the necessity of reconsidering human subjectivity in the modern era as radically different from its previous formations.

Various attempts have been made in the last two decades to use visualizations like these graphs to make an impact on governmental policy (and here I speak mainly of the United States and Western nations that have been, until recently, the primary culprits of greenhouse gas production). In 2008, for instance, the American environmentalist Bill McKibben founded 350.org, now an international organization dedicated to reducing carbon dioxide levels to 350 ppm—a number thought to be the tipping point at which CO_2 levels become unsafe for the planet. (For some comparison, when I first began research for this book [roughly in 2008], CO_2 levels were already at 385.45 ppm.[36] These levels have now risen [at the time of this writing, August 2013] to 395.15 ppm.[37]) Future projection models place CO_2 levels at the end of this century anywhere from 500 ppm to well above 900 ppm.[38] The self-destructive, linear direction of economic development, which shows absolutely no signs of weakening, promises to continue to force this worldwide hydrocarbon accident upon the planet's various ecosystems. Until recently, the United States was the major economic perpetrator of greenhouse gas production. China has now surpassed the United States in CO_2 emissions and is expected to produce double that of the United States by 2015.[39] This surge in greenhouse gases means that a human community is now contributing to global warming faster than any other country in history. The United States, however, should not become too comfortable. A different modeling shows that America still leads the world in terms of emissions per capita, at 17.3 tons per person per year (the world average is 4.5 tons), with Australia (17 tons), Russia (11.6 tons), Germany (9.3 tons), and the United Kingdom (7.8 tons) coming before China (5.4 tons).[40] In addition to China, India's emissions have also increased by 230 percent since 1990. This trend is expected to continue as well in other countries.

Resituating this knowledge against the critical concept of the human as a geological force can reveal its larger ideological constellation—namely, the complicity between the unsustainable practices of enclosing economic development, the unidirectional patterns of thought that bulldoze through more critically aware possibilities, the self-destructive closed-loop system of the Accidental, and the self-destructive zero-sum solutions proffered by the security society. As we saw in chapter 1, Cold War ideologies of security deemed it necessary to explode more than a thousand nuclear bombs in

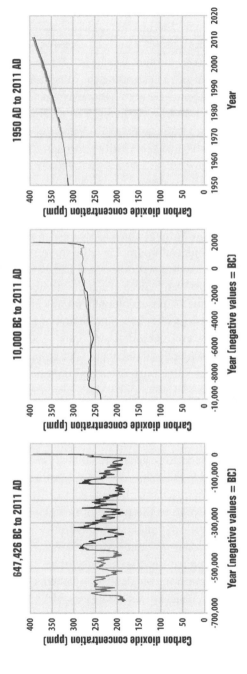

Figure 16. Global atmospheric concentrations of carbon dioxide given from three different temporal registers. Image courtesy of the Environmental Protection Agency.

the second half of the twentieth century, a program that encouraged a host of other nations to follow suit. The unidirectional narrative of the Prisoner's Dilemma offered logical "proof" that such activity was a necessary course of action—that we were, in other words, in a "state of exception." This coupling of an inherent good (defending the West against communism) to an exceptional state of necessity (that seeks to justify all courses of action) served as the logical driver that enforced a particular style of technological progress in the Cold War era. This double concealment—of an inherent good and an indisputable necessity—shrouds the vacuity of a supposedly organic technological endeavor. And it equally eclipses technology's collateral effects on ecosystems.

The toxic fallout from these Cold War experiments was long term and planet wide, affecting humans on a grand scale, and in this sense, they also share a similarity with the spatial character of global warming. Strontium 90, carbon 14, and plutonium 239—the three main radioactive isotopes produced in a nuclear reaction—have half-lives of 29 years, 5,700 years, and 24,000 years, respectively. Such are numbers not easily integrated into our accustomed ways of comprehending temporal reality. They are numbers closer in degree to the length of nonhuman geological ages. Radiation scientists Steven L. Simon, André Bouville, and Charles E. Land discovered that "by the early 1960s, there was no place on Earth where the signature of atmospheric nuclear testing could not be found in soil, water and even polar ice."[41] Unlike global warming, however, nuclear radiation more directly reveals its effects on the human body, and thus its toxic nature speaks immediately to an anthropocentric- and individual-oriented humanity. The worldwide spreading of radioactive pollutants, unlike climate change, immediately generated enough criticism in the United States to at least lead to the reduction of arms production between the United States and the Soviet Union. This was accomplished in part by directly linking atomic radiation to the spread of cancer, which became known publicly after antinuclear proliferation advocates Eric and Louise Reiss and Barry Commoner orchestrated the "Baby Tooth Survey" in the early 1960s. They pointed out, for instance, that "bone-seeking" strontium 90—the most dangerous element of radioactive fallout from a nuclear weapon—acts like calcium and moves toward bones, causing bone cancer and leukemia. Their study revealed that children born in 1963 had fifty times the level of strontium 90 in their teeth than children born before 1950. Increased levels of diseases also became known, caused by the absorption of the isotope into the bones. Simon, Bouville, and Land have more recently shown that there is "increasing proof that specific radionuclides in fallout are implicated in fallout-related cancers."[42] Arjun Makhijani's studies in the late 1990s estimated that the global radiation fallout from nuclear testing produced "hundreds of thousands of excess cancer fatalities."[43] In 1991 the International Physicians for the Prevention of Nuclear War released a study that estimated that radioactive fallout from atmospheric testing "would be responsible for 430,000 cancer deaths by the year 2000."[44] The study argued that an average of 2.4 million people could eventually die from the result of this testing.[45] Such numbers are even more staggering when we consider the unswerving course of atomic destruction adopted by scientists, military leaders, and politicians during World War II. The war machine was in this sense considerably incorrect in its conclusions about the "highly inefficient" yield of the bomb dropped on Hiroshima (which only released 1.38 percent of its possible total destructive power). Its "yield" hasn't even begun to be tallied.

At the same time that toxic fallout was spreading across the planet, apparatuses of security were becoming more global in their influence. It may be true, as Foucault has argued, that the age of biopower and security began in the eighteenth century, but the global range of security technologies is a phenomenon of the latter half of the twentieth century. Indeed, since the end of World War II, the U.S. national security state has expanded at a considerable rate. Defense spending in the United States has increased substantially since the early days of the Cold War. The Cold War period, 1947–89, ushered in an unprecedented level of military spending, with the United States and the USSR, not surprisingly, leading the world in military spending with 36 percent and 20 percent, respectively: "More economic resources were used for military purposes after World War II than ever before; during the 1980s the level of world military spending was more than 10 times higher than in the period 1925–38."[46] During the 1970s (and true to the Accidental logic of deploying insecurity), the U.S. Department of Defense exaggerated figures of Soviet military spending in order to justify increases in its own budget.[47] This misinformation was part of the basis for the Ford, Carter, and Reagan administrations to massively increase military spending from the 1970s to the 1980s, and it also led to a NATO general resolution that required its members to annually increase their military budgets by at least 3 percent each year. After the fall of the Soviet Union, world military spending decreased, but the United States, despite its own reductions in military spending (mostly a result of attempts to balance the budget), remained the world's largest military force throughout this period, spending upwards of $267 billion in 1997, which was more than 33 percent of total world military spending. In per-capita terms, this translates to a spending of roughly $1,000 per U.S. citizen, in relation to the world average of $240.[48] The attacks of 9/11 further exacerbated this spending. Since 2001, U.S. federal expenditures have increased $360 billion and state and local expenditures by $200 billion.[49] In 2004, world military spending exceeded $1 trillion; the United States accounted for 47 percent of that expenditure,[50] meaning that the U.S. budget exceeded the combined military spending of the next thirty-two countries.[51] More than a decade after 9/11, the United States remains the world's biggest military spender.

Not as widely acknowledged is the effect that 9/11 had on the expansion of the security society outside the United States. Developing nations that formerly had little or no security policies or strategies began to imbricate themselves in the expansion of a general planetary security society: "The new threat [of 9/11] provided a focus for national security strategies that had been lacking in most industrialized countries since the end of the cold war."[52] The United States' adoption of its new National Security Strategy paved the way for constant security measures and preemptive warfare based on the idea of "sufficient threat."[53] In addition to the U.S. development of preemptive warfare and security measures, 9/11 "helped to emphasize the reality of North-South interdependence in terms of security."[54] The European Union also began to highlight "nonmilitary" means for developing security mechanisms. In addition to the better-known national security initiatives of homeland security and the USA Patriot Act, in 2003 the Bush administration inaugurated a global security initiative designed to stop the trafficking of weapons of mass destruction. By 2012 that initiative was endorsed by 102 nations.[55] Known as the Proliferation Security Initiative, the law does not fall under any UN body, and its claim to its own self-established form of law enforcement has been disputed by China and India as violating the UN's Convention on the Law of the Sea. Even in the early years of post-9/11 security elaborations, the expansion of the global security network had already begun to move beyond Bush's presidential control of the state of exception and the sovereign-executive paradigm of power relations. According to Stockholm International Peace Research Institute (SIPRI) security analyst Allison Bailes, during the Iraq War, Western state security authorities were looking for ways to "maximize their 'export of security'" by "surrender[ing] larger parts of their sovereignty."[56] Four years after 9/11, the security society was seeking ways to interpellate "non-state as well as state actors in the developing world" to become a functioning part of a "comprehensive security solution."[57]

The expansion of the security society, the contamination of the planet from accidental toxic fallout, and the planetary nature of climate change are all dimensions of immense scale unknown to previous human centuries. To recall the scientist working for the Atomic Energy Commission who spoke of the "unbelievable" nature of world contamination brought about by the Pacific Proving Ground tests, "like Columbus when no one believed the world was round," these actions reveal the colossal nature of the particular version of the human that has become the dominant form of the species.[58] If we consider that the emissions causing global warming arise from the idea of enclosing production and development—the erasure, in other words, of inhabitancy—that informs the transformation of "Nature" into "Energy" (and the resultant need for "energy security"), then this singular transformation in human subjectivity into a geological force can be likened to the "one spot" in nuclear experimentation that "contaminates the world."

In his *Cultures of War*, John Dower effectively identifies a destructive imperative—a "terrible logic"—that seeks to expand the range and power of war's brute force to levels beyond previous and understandable proportions.[59] The first full-fledged appearance of this development, he argues, begins in World War II with the successful harnessing of nuclear power. But this "terrible logic" does not stop at the technoscientific level of production. The logic expands until it becomes a state of mind and general sociopolitical precept, spawning fixations of "deploying overwhelming forces" across an entire spectrum of the war machine's reach: in "power politics," in widespread technological development, and in a new "aestheticism of unrestrained violence" that influences cultural production as well.[60] As I have been arguing throughout this book, the spectrum of existence overcome by the lure of *immensity*—what Martin Heidegger once referred to as the movement of "the gigantic" from a quantitative to a qualitative status—has been expanding at an enormous and unchecked rate, generating the planetary gigantism of climate change.[61] As I've also argued, this technological and political search for ways to produce forces of immense proportionality began much earlier, with the rise of common law and the enclosure movement, a form of securitizing logic that subsequently informed military and technological developments in the nineteenth and twentieth centuries. These developments have set human subjectivity upon a particular path, and this path continues to affect even the most radical theoretical engagements with subjectivity to this day. With the erasure of inhabitancy comes the creation of a free-floating, "fluid" human subjectivity, capable of being displaced or made landless at any time at the whim of a large landowner, a large corporation, the unforeseeable twists of the global economic market, and now the unexpected events of environmental upheaval. As Edward Said wrote at the end of *Culture and Imperialism*, "Surely it is one of the unhappiest characteristics of the age to have produced more refugees, migrants, displaced persons, and exiles than ever before in history."[62] The erasure of inhabitancy places even those subjects living in domestic spaces in "nonlocalizable space," to use Donald Pease's term—the new spatiality favored by the "emergency state." As Pease argues, the "'Homeland' is that shifting imposition of placement without a place, of every subject being a potential terrorist and thus existing in a nonlocalizable state which makes capable the instant act of displacement and incarceration."[63] Now that "nonlocalizable" state of existence is in the process of being extended by the geological event of planetary climate change—making inhabitancy the void around which the postmodern security society is structured.

Chakrabarty's reimagining of humans as "geological forces" generating massive events of displacement is therefore a fresh critical concept, to say the least. It is an image of the human worth exploring at length. Westernized humanity has been increasingly relying on technologized patterns of thought governed teleologically by machines of war and security. The ideas generated by this form of intelligibility are too often taken for granted—key among these being the enclosing of land, the transformation of interdependent ecosystems into targets for the demands of war and energy security, and the growing momentum to proclaim adaptation as a primary strategy for addressing anthropogenic climate change. These ideas in turn produce material structures of immense size and impact. All this apparently indispensable or "exceptional" activity transforms consciousness, to invoke Adorno, into more of an "industry" than a creative force for change. Once transformed into an industry, consciousness can do little else than generate more of the same, leading to the asphyxiating reduction of transcendent moments into accidental fallout. Being dependent on these "efficient" technologically advanced spectacles for our existence makes it all the more difficult to challenge their predominance. Such addictions to the spectacular make it difficult to unconceal their affiliation with the coercive demands imposed on human existence by corporations, security institutions, military leaders, government administrators, and security specialists. The industrial mind revels in responding to these demands by producing increasingly larger technological marvels, structures that replace the otherworldly possibilities of the sublime with the calculative confinements of the beautiful. At such moments we can begin to see the uncanny similarity between the collateral fallout of nuclear testing and the collateral displacement and ecological devastation brought about by modern marvels like the Belo Monte megadam. And the Amazon is not the only ecosystem threatened by such projects: similar megadam projects are under way in Malaysia, Ethiopia, and China. Even the United States plans to bring back the era of the megadam. Proposals are under way to build the Hoover-sized Susitna dam in Alaska. There are countless similar proposals of technological immensity, as we saw in Robert Zubrin's idea to transform a large swath of Latin America into a sugarcane farm for the continuation and expansion of the U.S. automobile industry. Geoengineering, hydraulic fracking, offshore oil drilling—these are structures of increasingly immense size and impact, ensuring the continuation of the animality of man and its unrelenting velocity. These structures in turn are buttressed by metaphysical modes of representation, as we saw in chapter 1, that almost entirely erase alternative concepts, efforts, and struggles.

Again, these technological marvels cannot be thought of apart from the transformation of the human into a geological force, a transformation that is, in turn, made possible by the collapse of the division between Man and Animal. At a certain point in his analysis of this animality of Man, Agamben suggests that such a collapse may have been made possible by the twentieth century's most violent manifestation of the state of exception: "When the difference vanishes and the two terms collapse upon each other—as seems to be happening today—the difference between being and the nothing, licit and illicit, divine and demonic also fades away, and in its place something appears for which we seem to lack even a name. Perhaps concentration and extermination camps are also an experiment of this sort, an extreme and monstrous attempt to decide between the human and the inhuman, which has ended up dragging the very possibility of the distinction to its ruin."[64] Agamben suggests here that the act of a finalized division, or more precisely the collapse of the human–animal division, shares an affinity with the twentieth century's monstrous acts of war and enclosure. The brutal incarceration and relentless slaughter of entire races signifies the final attempt to put an end to the question, thus destroying any possibility of bringing the distinction back into existence. Perhaps too this monstrous act was a form of adaptation—that is, of adapting to the threat of limited resources through the annihilation of an entire race, a race understood to be, for all intents and purposes, another species. Or, put differently, this monstrous act was an attempt to discover an outside cause for capitalism's innate production of insecurity, justifying a massive system of security that eventually had to destroy millions to continue its existence. In the age of climate change, this generation of insecurity achieves its strongest actualization, spreading "natural security's" battle for life toward anyone and everyone and toward everything—resulting in what will most likely end up being an "adaptation of the fittest." At such a point—when the boundaries of the human are no longer questionable and the human can only claim the right to wage war over life—we can say without question that the age of environmentality has come to rule.

But at such point too—when securitizing forces, moving beyond their previous proportions, take on a geological quality—the accidental breakdowns that follow paradoxically bring to the foreground, more and more, the specter of inhabitancy. The growing political demand for inhabitancy, now planet wide, is not a fantasy. It constitutes a struggle that directly confronts today's ecological exigencies and names a form of existence that arrives from outside the rule of environmentality. Such activities constitute the power of the political as identified by Rancière: they are performances that uncover the inequality of geological forces bearing down upon us all—the "above-surface structures" that install, like Kipling's Kim, a generalized fear and suspicion of the environment. The inhabitants of the twenty-first century are the shadow humans (and nonhumans) that haunt the defensive SAGE palaces of the Global North and West. And they offer an audacious exhortation to the forces of environmentality governing reality in the age of climate change.

Notes

Introduction

1. Sharon Burke and Christine Parthemore, "Climate Change War Game: Major Findings and Background" (working paper, Center for a New American Security, Washington, D.C., June 2009), 6.

2. Ibid., 8.

3. For a history of the early 1960s environmental movement in the United States and the work of Carson, Leopold, and Commoner, see Peter Hay, *Main Currents in Western Environmental Thought* (Bloomington: Indiana University Press, 2002), 15–20.

4. Raphael D. Sagarin and Terence Taylor, *Natural Security: A Darwinian Approach to a Dangerous World* (Berkeley: University of California Press, 2008), 12.

5. Sagarin, *Natural Security*, front cover flap.

6. "Operational Adaptation," http://www.operationaladaptation.com.

7. Sagarin, *Natural Security*, xiv–xv.

8. This is from Gary Hart's introduction to Rafe Sagarin's *Learning from the Octopus: How Secrets from Nature Can Help Us Fight Terrorist Attacks, Natural Disasters, and Disease* (New York: Basic Books, 2012), xv.

9. Amanda Griscom Little, "Who's Afraid of the Big, Bad Woolsey?," *Grist*, June 7, 2005, http://www.grist.org.

10. See the official Green Patriot website, http://www.greenpatriot.us.

11. Douglas Jehl, "C.I.A. Nominee Wary of Budget Cuts," *New York Times*, February 3, 1993, http://www.nytimes.com.

12. The CIA announced as this book was going to press that it was shutting down MEDEA and its secret climate research program. See Tim McDonell, "Exclusive: The CIA Is Shuttering a Secretive Climate Research Program," *Mother Jones*, May 21, 2015, http://www.motherjones.com/environment/2015/05/cia-closing-its-main-climate-research-program.

13. Carolyn Pumphrey, ed., *Global Climate Change: National Security Implications* (Carlisle, Pa.: Strategic Studies Institute, Army War College, 2008), http://www.strategicstudiesinstitute.army.mil/pdffiles/pub862.pdf.

14. Rymn J. Parsons, "Taking Up the Security Challenge of Climate Change" (Carlisle, Pa.: Strategic Studies Institute, August 2009).

15. Ibid., 1.

16. Ibid., 2.

17. U.S. Department of Defense, *Quadrennial Defense Review Report*, February 2010, http://www.defense.gov/qdr/images/QDR_as_of_12Feb10_1000.pdf.

18. David Nakamura, "At Coast Guard Graduation, Obama Warns of Climate Change Threat to National Security," *Washington Post*, May 20, 2015.

19. The Center for Naval Analysis Corporation, *National Security and the Threat of Climate Change*, 2007, 10, https://www.cna.org.

20. Funds for adaptation increased from $2.3 billion to $4.6 billion. Funds for mitigation increased from $7 billion to $7.1 billion. "World Bank Lending Doubles on Adaptation," *The World Bank: News*, November 15, 2012, http://www.worldbank.org/en/news/feature/2012/11/15/World-bank-lending-doubles-adaptation.

21. The World Bank, "Climate Change Adaptation," http://www.worldbank.org/en/topic/climatechange.

22. The International Monetary Fund, "Climate, the Environment, and the IMF," http://www.imf.org; World Trade Organization, "Activities of the WTO and the Challenge of Climate Change," http://www.wto.org.

23. Elizabeth DeLoughrey, "Radiation Ecologies and the Wars of Light," *Modern Fiction Studies* 53, no. 3 (Fall 2009): 468–95.

24. Tobias Smollett, *The Expedition of Humphrey Clinker* (New York: Penguin, 2008), 285.

25. Mary Louise Pratt, *Imperial Eyes: Travel Writing and Transculturation* (New York: Routledge, 2008), 34.

26. Ibid., 34.

27. Stephen Koblik names Linnaean scientific inquiry and the enclosure movement as "the most critical changes in Swedish history." See Stephen Koblik, ed., *Sweden's Development from Poverty to Affluence, 1750–1970* (New Berlin, Wis.: University of Minnesota Press, 1975), 12.

28. Pratt, *Imperial Eyes*, 35.

29. Rob Nixon, *Slow Violence and the Environmentalism of the Poor* (Cambridge, Mass.: Harvard University Press, 2011), 41.

30. Jacob Darwin Hamblin, *Arming Mother Nature: The Birth of Catastrophic Environmentalism* (Oxford: Oxford University Press, 2013), 135.

31. Ibid., 135.

32. Ibid., 136–37.

33. See Daniel Knight, "Rights—Nigeria: Oil Giant Had Role in Killings," *Inter Press Service*, October 2, 1998, http://www.ipsnews.net/1998/10/rights-nigeria-oil-giant-had-role-in-killings; Human Rights Watch, *The Price of Oil: Corporate Responsibility and Human Rights Violations in Nigeria's Oil Producing Communities*

(New York: Human Rights Watch, 1999), http://www.hrw.org/reports/1999/
nigeria/nigeria0199.pdf.

34. Dan Stillman, "DoD Takes Aim at Climate Change," *Imaging Notes*, Spring
2010, http://www.imagingnotes.com/go/article.php?mp_id=217.

35. Giorgio Agamben, *Homo Sacer: Sovereign Power and Bare Life* (Stanford,
Calif.: Stanford University Press, 1998), 20.

36. Although different polling agencies tell slightly different stories, the gen-
eral trend that Republicans are more likely to deny climate change or its effects
holds. For recent Gallup polls, see "In U.S., Global Warming Views Steady Despite
Warm Winter," http://www.gallup.com/poll/153608; "Americans' Worries about
Global Warming Up Slightly," http://www.gallup.com/poll/153653; "Conservatives'
Doubts about Global Warming Grow," http://www.gallup.com/poll/126563. For an
alternative, see the Associate Press poll: "AP-GfK Poll: Belief in Global Warming
Rises with Thermometers, Even among US Science Doubters," http://ap-gfkpoll
.com/uncatergorized/out-latest-poll-findings-18.

37. Sean Pool, "House Energy and Commerce Committee Votes for Science
Denial," *Science Progress*, March 16, 2011, http://scienceprogress.org/2011/03/house
-energy-and-commerce-committee-votes-for-science-denial.

38. Dina Fine Maron, "New Anti-EPA Bill Aims to 'Rein in' Agency's Climate
Rules Permanently," *New York Times*, March 4, 2011, http://www.nytimes.com.

39. Brad Johnson, "South Dakota Legislatures Tell Schools to Teach 'Astrolog-
ical' Explanation for Global Warming," *Think Progress*, February 25, 2012, http://
thinkprogress.org/politics/2010/02/25/83917.

1. The SAGEs of the Earth and the Accidental Nature of Environmentality

1. I am indebted to Elizabeth DeLoughrey for bringing this to my attention.

2. The U.S. Department of Energy, "Albert Einstein to Franklin D. Roosevelt—
August 2, 1939," *The Manhattan Project: An Interactive History*, https://www.osti.gov/
opennet/manhattan-project-history/Resources/einstein_letter_photograph.htm#1.

3. Alice Kimball Smith and Charles Weiner, eds., *Robert Oppenheimer: Let-
ters and Recollections* (Cambridge, Mass.: Harvard University Press, 1980), 317.
Quoted in John W. Dower, *Cultures of War: Pearl Harbor/Hiroshima/9/11/Iraq*
(New York: W. W. Norton, 2010), 256.

4. Ibid., xxi–xxii.

5. Jacob Darwin Hamblin, *Arming Mother Nature: The Birth of Catastrophic
Environmentalism* (Oxford: Oxford University Press, 2013), 38.

6. Ibid., 39.

7. Robert K. Plumb, "Man-Made Threat to Weather Seen: U.S. Scientist Says
Studies Indicate World Climate Is Subject to Control," *New York Times*, January 28, 1958, 54.

8. Paul Virilio, *Unknown Quantity* (London: Thames and Hudson, 2003), 6.

9. Elizabeth DeLoughrey was the first to bring this event to my attention. See her "Radiation Ecologies and the Wars of Light," *Modern Fiction Studies* 53, no. 3 (Fall 2009): 468–95. This excellent essay theorizes the consequences of a "fully enlightened earth" and how light radiation becomes an asphyxiating, planetary phenomenon.

10. Ibid., 474.

11. Ibid., 491.

12. Ibid., 491.

13. United Nations, *Report of the United Nations Scientific Committee on the Effects of Atomic Radiation to the General Assembly*, 2000, 5.

14. Comprehensive Nuclear-Test-Ban Treaty Organization, "General Overview of the Effects of Nuclear Testing," 2012, http://www.ctbto.org.

15. Susan Solomon, Gian-Kasper Plattner, Reto Knutti, and Pierre Friedlingstein, "Irreversible Climate Change Due to Carbon Dioxide Emissions," *PNAS* (*Proceedings of the National Academy of Sciences of the United States of America*) 106, no. 6 (2009): 1704–9, 1704.

16. Slavoj Žižek, *The Fragile Absolute: Or, Why Is the Christian Legacy Worth Fighting For?* (London: Verso, 2000), 15.

17. Lauren Morello, "Defense Experts Press for Probabilistic Risk Assessment of Climate Change," *Scientific American*, June 24, 2010, http://www.scientificamerican.com.

18. Ibid.

19. Ibid.

20. For a discussion of the concrete universal, see Charles Taylor, *Hegel* (Cambridge: Cambridge University Press, 1975), esp. 112–14. See also Žižek's distinct development of the concrete universal, to which my own work is more aligned, in *The Ticklish Subject: The Absent Centre of Political Ontology* (London: Verso, 1999), esp. 177–228.

21. Barry Comoner, *Science and Survival* (New York: Viking, 1967), 25.

22. Ibid., 23.

23. Paul Virilio, *War and Cinema: The Logistics of Perception* (London: Verso, 1989), esp. 1–6.

24. Ibid., 2.

25. Manuel De Landa, *War in the Age of Intelligent Machines* (New York: Zone Books, 1991), 22.

26. Ibid., 34.

27. Judith Butler, "Contingent Foundations: Feminism and the Question of 'Postmodernism,'" in *Feminists Theorize the Political*, ed. Judith Butler and Joan W. Scott (New York: Routledge, 1992), 11.

28. James M. Garrett, *Wordsworth and the Writing of the Nation* (London: Ashgate, 2008), 70.

29. Mark Monmonier, "The Rise of Map Use by Elite Newspapers in England, Canada, and the United States," *Imago Mundi* 38 (1986): 46–60.

30. Matthew H. Edney, *Mapping an Empire: The Geographical Construction of British India, 1765–1843* (Chicago: University of Chicago Press, 1999), 33.

31. Ibid., 31.

32. "It is here that the apparent perfection of the geographical panopticon promised by the Great Trigonometrical Survey is revealed as an empiricist delusion. The chaotic circumstances of British surveying in India are conveniently obscured by a veneer of order and system. The ordered hierarchy of dependence between the surveys did not in fact occur. Whatever order was imposed on the detailed surveys came *after* the fact, when they were incorporated into the general archive. The order did not derive directly from the surveys themselves. What the British implemented was not the ideal, but only the *image* of the ideal." Ibid., 30.

33. In Martin Heidegger's sense of the Romanization of the ancient Greek term *meta-ta-physika*, meaning "above, beyond, outside" context, flux, particularity, immanence, and so on. To be in a metaphysical position proposes that one has risen above the fray and achieved a superior vantage point—a final or unquestionably objective stance that has overcome the limitations of a bounded, immediate context. "Keeping one's distance" from regional contamination is one of the prerequisites of achieving a metaphysical viewpoint. It is the ontology of metaphysics, for example, that informs popular neoliberal and American discourses such as "the end of history" and "American exceptionalism." See Martin Heidegger, "What Is Metaphysics?," in *Basic Writings*, ed. and trans. David Farrell Krell (New York: Harper and Row, 1977), 109.

34. A brief list would include Richard Schacht, *Nietzsche* (London: Routledge, 1983), esp. 118–86; Heidegger, "What Is Metaphysics?," esp. 106–7; Martin Heidegger, "The Onto-theo-logical Constitution of Metaphysics," in *Identity and Difference*, trans. Joan Stambaugh (Chicago: University of Chicago Press, 1969), 42–73; Jacques Derrida, *Dissemination*, trans. Barbara Johnson (Chicago: University of Chicago Press, 1981), esp. 181–226 and also Johnson's "Translator's Introduction," esp. viii–x; Michel Foucault, *The Courage of Truth, Lectures at the Collège de France, 1982–1983*, trans. Graham Burchell (Basingstoke: Palgrave Macmillan, 2011), esp. 236–37; Julia Kristeva, *Black Sun: Depression and Melancholia*, trans. Leon S. Roudiez (New York: Columbia University Press, 1992), esp. 66–67; Gayatri Chakravorty Spivak, *The Post-Colonial Critic: Interviews, Strategies, Dialogues* (London: Routledge, 1990), esp. 7.

35. Heidegger, "What Is Metaphysics?," 106–7. Heidegger argues that critique involves a necessary metaphysical gesture in order to see the limitations of what appears by rising above those limitations in a gesture that reveals the structures governing the *ways in which* things appear. The Westernization of the *meta-ta-physika*, however, transforms this critical gesture into an act of ontological imperialism, whereby transcendence becomes the imposition of an

essential positivism or identity (on a race, culture, or ecosystem, for instance)—
what Derrida would identify as the "principle of presence." William V. Spanos
has developed these connections at length: "To emphasize its visual orientation,
inquiry understood as adequation of mind to thing proceeds from *above* (*meta-ta-
physika*): from a fixed transcendental vantage point—a 'Transcendental Signified'
or 'center elsewhere,' in Derrida's terms, which is beyond the reach of free play.
In thus privileging the surveying and globalizing eye of vision, it has its ultimate
purpose in the coercion of difference into the circumference of the identical circle.
The center spatializes and reifies the disseminations of temporality in order to
'comprehend' them: not simply to know but to 'take hold of' or 'manage,' that is,
dominate and use them. The comportment toward phenomena the center enables
is thus that of the *commanding* eye, that is to say, the panoptic gaze. It is a visual
comportment that represents the force of difference as that which truth is not, as
false (*falsum*) and thus as a threat to truth that must be domesticated or pacified,
willfully reduced to the same in the name of justice." See Spanos, *Heidegger and
Criticism: Retrieving the Cultural Politics of Destruction* (Minneapolis: University
of Minnesota Press, 1993), 142.

36. On C3i, see Paul Virilio, *War and Cinema: The Logistics of Perception*
(London: Verso, 1989), 2. On the "higher-order motor," see Manuel De Landa, *War
in the Age of Intelligent Machines*, 4. The word order is sometimes changed in the
military literature to "communication, command, control, intelligence."

37. Rudyard Kipling, *Kim* (New York: Penguin, 1989), 164.

38. Ibid., 337–38.

39. Hamblin, *Arming Mother Nature*, 173.

40. Paul N. Edwards, *The Closed World: Computers and the Politics of Discourse in Cold War America* (Cambridge, Mass.: MIT Press, 1997), 75.

41. This was the designation given to Bush by *Time* magazine in 1944. See
Peter Dizikes, "The General of Physics," *Technology Review*, April 19, 2011, http://
www.technologyreview.com/mitnews/423694/the-general-of-physics.

42. Eric G. Swedin and David L. Ferro, *Computers: The Life Story of a Technology* (Baltimore: Johns Hopkins University Press, 2007), 49.

43. Edwards, *The Closed World*, 75, 101.

44. "Sage Air Defense," http://homepage.mac.com/oldtownman/20th/sage
.html.

45. Edwards, *The Closed World*, 1.

46. Paul Edwards, "Computers and Weapon Systems," in *Computers in Battle: Will They Work?*, ed. David Bellin and Gary Chapman (New York: Harcourt,
Brace, Jovanovich, 1987), 54. Quoted in De Landa, *War in the Age*, 54.

47. The Center for Naval Analysis Corporation, *National Security and the
Threat of Climate Change*, 2007, 10, https://www.cna.org.

48. Edwards, *The Closed World*, 7.

49. This phrase is placed in quotations so as to reference the term *C3i*: the command, control, and communication of information.

50. William J. Broad, "C.I.A. Is Sharing Data with Climate Scientists," *New York Times*, January 4, 2010.

51. Ann Finkbeiner, *The Jasons: The Secret History of Science's Postwar Elite* (New York: Viking, 2000).

52. Giorgio Agamben, "Marginal Notes on *Commentaries on the Society of the Spectacle*," in *Means without End: Notes on Politics*, trans. V. Binetti and C. Casarino (Minneapolis: University of Minnesota Press, 2000).

53. I am indebted to William V. Spanos for bringing this essay by Agamben to my attention.

54. Rob Nixon, *Slow Violence and the Environmentalism of the Poor* (Cambridge, Mass.: Harvard University Press, 2011), 2–3.

55. Louis Althusser develops the notion of the problematic as an ideological framework that circumscribes and delineates "the forms in which problems must be posed." It predetermines the questions that an inquirer can ask and consequently determines the very answers these questions produce: "[Science] can only pose problems on the terrain and within the horizon of a definite theoretical structure: its problematic, which constitutes its absolute and definite condition of possibility, and hence the absolute determination of *the forms in which all problems must be posed*, at any given moment in the science." In terms of contemporary ideological processes of image making, Althusser's conception here is particularly useful in that it foregrounds the relation between problem posing and the visual: "Any object or problem . . . in the definite structured field of the theoretical problematic of a given discipline . . . is visible. . . . The sighting is thus no longer the act of an individual subject, endowed with the faculty of 'vision' which he exercises either attentively or distractedly; the sighting is the act of its structural conditions. . . . It is literally no longer the eye of a subject which *sees* what exists in the field defined by a theoretical problematic: it is the field which *sees itself* in the objects of problems it defines—sighting being merely the necessary reflection of the field on its objects. . . . The same connexion that defines the visible also defines the invisible as its shadowy observe. It is the field of the problematic that defines and structures the invisible as the defined excluded, *excluded* from the field of visibility and *defined* as excluded by the existence and peculiar structure of the field of the problematic. . . . These new objects and problems are necessarily *invisible* in the existing theory, because they are not objects of this theory, because they are *forbidden* by it—they are objects and problems necessarily without any necessary relations with the field of the visible as defined by the problematic. They are invisible because they are rejected in principle, repressed from the field of the visible; and that is why their fleeting presence in the field when it does occur (in very peculiar and symptomatic circumstances) *goes unperceived,* and becomes literally

an undivulged absence—since the whole function of the field is not to see them, to forbid any sighting of them. Here again, the invisible is no more a function of *a subject's sighting* than is the visible: the invisible is the theoretical problematic's non-vision of its non-objects, the invisible is the darkness, the blinded eye of the theoretical problematic's self-reflection when it scans its non-objects, its non-problems without seeing them, *in order not to look at them.*" Louis Althusser, *Reading Capital* (London: Verso, 1979), 25–26.

56. Miguel Llanos, "60 Dams in Brazil's Amazon? Controversy Spills over into 'Earth Summit II,'" http://worldnews.nbcnews.com.

57. For the most part, the relation between the military and the environment has received scattered attention by ecocritics, despite the expansion of the field of ecocriticism into global and postcolonial concerns—two disciplines that have an established history of addressing military activities. Two notable exceptions can be found in the recent work of Elizabeth DeLoughrey and Mike Hill. See especially DeLoughrey's "Radiation Ecologies and the Wars of Light," *Modern Fiction Studies* 55, no. 3 (Fall 2009): 468–95; and Hill's "Ecologies of War: Dispatch from the Aerial Empire," in *Telemorphosis: Theory in the Era of Climate Change*, vol. 1, ed. Tom Cohen (Ann Arbor, Mich.: Open University Press, 2012), 239–69.

58. Paul Virilio, *Popular Defense and Ecological Struggles* (New York: Semiotext[e], 1978), 65–66.

59. Commentary attached to these images states, "These maps show warming of the Earth in 2050 and 2100 relative to the beginning of the 21st century. Darker coloring represents greater warming. The data were produced by the Community Climate System Model, version 3 (CCSM3) developed by the National Center for Atmospheric Research in the United States. The data were processed and the maps were produced by the Oak Ridge National Laboratory. The greenhouse gas emissions scenario driving the model was what is commonly called 'business as usual' emissions." This description implies that emissions continue to grow based on rapid economic growth and a fossil fuel–intensive energy system, in the absence of policies to limit greenhouse gas emissions in the future. In this scenario, CO_2 concentration in the atmosphere is between 450 and 500 ppm (parts per million) in 2050 and between 900 and 1000 ppm in 2100. "As the Secretary general pointed out, there is little we can do to alter the climate of 2050 because it will be governed by the greenhouse gases we have already emitted. The 2012 agreement would limit CO_2 concentrations to around 450 ppm, which would produce a stable climate slightly cooler than the one depicted in the upper map for 2050. Unfortunately, the IPCC estimates that we would have needed our global CO_2 emissions to peak this year in order to achieve that goal, yet our emissions continue to grow rapidly. On average this world is about 2.5 degrees warmer than the preindustrial world and some regions near the poles warm by about 6 degrees. This is a challenging world to live in

because hundreds of millions to billions more people will live under water stress, weather will be significantly more extreme, and some diseases are likely to spread to new areas. The developing world will be impacted the most, especially Africa, Asia, and tropical Latin America. But if we stabilize the climate in this condition, there is a lot we can do to adapt to those changes and there is a chance that we can avoid the most severe impacts of climate change, such as multiple meters of sea level rise before the end of this century and the collapse of the North Atlantic Ocean's conveyor belt or thermohaline circulation, which would likely have dramatic effects for the global climate, not just in Europe. . . . The lower map shows a very different world that is almost 6 degrees warmer on average. This is a world that humans have never known. The Earth has not been this warm for more than 50 million years. The last time it was this warm there was no ice on Greenland or Antarctica and the global average sea level was 90 meters higher than it is today. Permanent ice cannot survive on a planet this warm. And although it would take thousands of years for all of the ice to melt, only small portions of the ice are needed to cause catastrophic sea level rise on the scale of a century. A world this warm is very likely to undergo other large, abrupt changes, such as reorganization of the atmosphere that shifts rainfall away from our main food producing regions and the collapse of the North Atlantic thermohaline circulation. Such a climate would likely result in fundamental reorganization of the global society in ways that we simply cannot assess or prepare for. And it is very important to understand also that CO_2-driven warming is essentially irreversible on any time scale that humans care about. Whether we allow the planet to warm by two degrees or six degrees, to the best of our knowledge we will be stuck with that temperature for thousands of years." The Center for a New American Security, "Climate Change War Game: Angry Red Future Slides," http://www.cnas.org/files/uploads/ angry_red_future_slide_podesta.pdf.

60. Rob Nixon, "Environmentalism and Postcolonialism," in *Postcolonial Studies and Beyond*, ed. Ania Loomba, Suvir Kaul, Matti Bunzl, Antoinette Burton, and Jed Esty (Durham, N.C.: Duke University Press, 2005), 233–51.

61. Figures 11 and 12 were orchestrated with the help of aerial artist John Quigley. See his website, suggestively called "Spectral Q: Collaborative Art for the Common Good," http://www.spectralq.com/Home.html.

62. For more information on the activity of the ABIP, see the indigenous publication *Earth Peoples*, http://earthpeoples.org/blog/?p=1472.

63. http://abcnews.go.com/blogs/headling/2012/06.

64. As far as I have been able to discover, out of the major networks, only NBC published the "Pare Belo Monte" image. CNN had covered protests of the project in the past, but no record of any coverage of the events of June is to be found on their website. Fox News reproduced no pictures of any of the protests. Instead, they ran a story about Brazilian tribal leaders holding Belo Monte engineers

hostage. The images appeared in several alternative sources, such as the U.S.-based Amazon Watch organization: http://amazonwatch.org/news/2012.

65. http://news.mongabay.com.

66. See Jacques Rancière, *Disagreement: Politics and Philosophy* (Minneapolis: University of Minnesota Press, 1999). "Docile bodies" is, of course, Foucault's definition for the subject that tacitly consents to established flows of power, the subject that is both "economically useful" but "politically obedient." See Michel Foucault, *Discipline and Punish: The Birth of the Prison* (New York: Vintage Books, 1979), 141. "Standing-reserve" is Heidegger's term for the subject that consents to step into the "world picture"—the particular ideological structure that holds sway over and defines human existence. See Martin Heidegger, "The Age of the World Picture," in *The Question Concerning Technology and Other Essays* (New York: HarperCollins, 1982).

67. *Environmentality* can broadly be defined as a momentum orchestrated within the discourse of security studies to politically address the environmental and geopolitical consequences of climate change. It refers to an essential and growing connection among surveillance technologies, security discourses, the military, global neoliberal expansion, neoconservative concerns for terrorism and national security, and the environment. For an expanded discussion of this, see my essay "Energy Security: The Planetary Fulfillment of the Enclosure Movement," *Radical History Review* 109 (Fall 2009): 83–99.

2. Inhabitancy, Custom Law, and the Landless

1. Perry Miller, *Errand into the Wilderness* (Cambridge, Mass.: Harvard University Press, 1956); Sacvan Bercovitch, *The American Jeremiad* (Madison: University of Wisconsin Press, 1978); Donald Pease, "New Americanists: Revisionist Interventions into the Canon," in "*New* Americanists," special issue, *boundary 2* 17, no. 1 (Spring 1990); Donald Pease, *The New American Exceptionalism* (Minneapolis: University of Minnesota Press, 2009); William V. Spanos, *American Exceptionalism in the Age of Globalization: The Specter of Vietnam* (Albany: State University of New York Press, 2008), esp. 187–241.

2. Genesis 1:26. For a discussion of Bacon's scientific method and its relation to nature, see Carolyn Merchant, *The Death of Nature: Women, Ecology and the Scientific Revolution* (New York: HarperCollins, 1980), esp. 164–77.

3. Helge Kragh, *Quantum Generations: A History of Physics in the Twentieth Century* (Princeton, N.J.: Princeton University Press, 1999), 400.

4. Paul N. Edwards, *The Closed World: Computers and the Politics of Discourse in Cold War America* (Cambridge, Mass.: MIT Press, 1997), 8.

5. Robert Zubrin, *Energy Victory: Winning the War on Terror by Breaking Free of Oil* (Amherst, N.Y.: Prometheus Books, 2007), 259–60.

6. Ted Nordhaus and Michael Shellenberger, *Break Through: From the Death of Environmentalism to the Politics of Possibility* (Boston, Mass.: Houghton Mifflin Harcourt, 2007); Thomas Friedman, *Longitudes and Latitudes: The World in the Age of Terrorism* (New York: Anchor Books, 2003); Thomas Friedman, *The World Is Flat: A Brief History of the Twenty-First Century* (New York: Farrar, Straus, and Giroux, 2005); Thomas Friedman, *Hot, Flat, and Crowded: Why We Need a Green Revolution—and How It Can Renew America* (New York: Farrar, Straus, and Giroux, 2008).

7. E. P. Thompson, *The Making of the English Working Class* (New York: Penguin, 1991), 237.

8. See, for instance, Michael Goldman, ed., *Privatizing Nature: Political Struggles for the Global Commons* (New Brunswick, N.J.: Rutgers University Press, 1998); Kathryn Harrison and Lisa McIntosh Sundstrom, *Global Commons, Domestic Decisions: The Comparative Politics of Climate Change* (Cambridge, Mass.: MIT Press, 2010); Susan J. Buck, *The Global Commons: An Introduction* (Washington, D.C.: Island, 1998); William D. Nordhaus, *Managing the Global Commons: The Economics of Climate Change* (Cambridge, Mass.: MIT Press, 1994); Vandana Shiva, *Biopiracy: The Plunder of Nature and Knowledge* (Boston: South End, 1987); Burns H. Weston and David Bollier, *Green Governance: Ecological Survival, Human Rights, and the Law of the Commons* (Cambridge: Cambridge University Press, 2013). Jeremy Rifkin perhaps comes closest to discussing the relationship of the commons to the specifics of the enclosure movement, but even his short treatment of the specifics of the history of enclosure is limited. See Jeremy Rifkin, *Biosphere Politics: A New Consciousness for a New Century* (New York: Crown, 1991). The brutalities and contradictions of enclosure are particularly overlooked by those influenced by the individualist-oriented philosophy of Garrett Hardin, who saw the commons as a "tragedy." See the essays collected in John A. Baden and Douglass S. Noonan, eds., *Managing the Commons* (Indianapolis: Indiana University Press, 1998).

9. Garret Hardin, "The Tragedy of the Commons," *Science* 162, no. 3859 (1968): 1243–48.

10. Gordon Batho, "Landlords in England," in *The Agrarian History of England and Wales*, vol. 4, *1500–1640* (Cambridge: Cambridge University Press, 1967), 256; for a concise description of the common field, see Joan Thirsk, "The Common Fields," in *Peasants, Knights, and Heretics: Studies in Medieval English Social History*, ed. R. H. Hilton (Cambridge: Cambridge University Press, 1981), esp. 10–11; for a discussion of "mead sticks," see W. E. Tate, *The English Village Community and the Enclosure Movement* (London: Victor Gollancz, 1967), 97.

11. For the specifics of feudal social hierarchy, see Richard Lachmann, *From Manor to Market: Structural Change in England, 1536–1640* (Madison: University of Wisconsin Press, 1987).

12. Elinor Ostrom, *Governing the Commons: The Evolution of Institutions for Collective Action* (Cambridge: Cambridge University Press, 1990).

13. See, for instance, Jeremy Rifkin, *Biosphere Politics: A New Consciousness for a New Age* (New York: Crown, 1991), esp. 37–38, 71–80; and "New Enclosures," special issue, *Radical History Review* 109 (Winter 2011).

14. "Managing and Owning the Landscape," http://www.parliament.uk/about/livingheritage.

15. W. E. Tate, *The English Village Community and the Enclosure Movements* (London: Victor Gollancz, 1967), 124.

16. Ibid., 140.

17. A brief list includes, in America: Allan Greer, "Commons and Enclosure in the Colonization of North America," *American Historical Review* 117, no. 2 (2012): 365–86; Peter Linebaugh and Marcus Rediker, *The Many-Headed Hydra: Sailors, Slaves, Commoners and the Hidden History of the Revolutionary Atlantic* (Boston: Beacon, 2000); Francis Jennings, *The Invasion of America: Indians, Colonialism, and the Cant of Conquest* (Chapel Hill: University of North Carolina Press, 2010), 82–84; Mexico: John Womack, *Zapata and the Mexican Revolution* (New York: Random House, 1969), 41–47, 50–54; India: Madhav Gadbil and Ramachandra Guha, *This Fissured Land: An Ecological History of India* (Berkeley: University of California Press, 1993), 34–35, 53–59, 115–16; Brazil: Gabriel Ondetti, *Land, Protest, and Politics: The Landless Movement and the Struggle for Agrarian Reform in Brazil* (University Park: Pennsylvania State University Press, 2008), 26–28; Africa: Daniel W. Bromley, "The Enclosure Movement Revisited: The South African Commons," *Journal of Economic Issues* 28, no. 2 (June 1994): 357–65; Lance van Sittert, "Holding the Line: The Rural Enclosure Movement in the Cape Colony, c. 1865–1910," *Journal of African History* 43, no. 1 (2002): 95–118. For a history of sugarcane production in Brazil, see Stuart B. Schwartz, *Sugar Plantations in the Formation of Brazilian Society: Bahia, 1550–1835* (Cambridge: Cambridge University Press, 1985); Thomas D. Rogers, *The Deepest Wounds: A Labor and Environmental History of Sugar in Northeast Brazil* (Chapel Hill: University of North Carolina Press, 2010).

18. Jean-Phillippe Platteau, "The Evolutionary Theory of Land Rights as Applied to Sub-Saharan Africa: A Critical Assessment," *Development and Change* 27 (1996): 29–86.

19. George Caffentzis, *In Letters of Blood and Fire: Work, Machines, and the Crisis of Capitalism* (Brooklyn, N.Y.: Autonomedia, 2013), esp. 236–40.

20. Stuart Piggott, H. E. Hallam, Joan Thirsk, Edward Miller, E. J. T. Collins, G. E. Mingay, and Edith H. Whetham, eds., *The Agrarian History of England and Wales* (Cambridge: Cambridge University Press, 1967, 1985, 2011). Symptomatically, it is during the twentieth century—the century when the former colonies achieved (relative) independence—that the movement entered the academy as a

highly contested discipline of study. Max Weber, one of the founders of sociology, takes on a study of ancient agricultural methods and relations before the publication of *The Protestant Ethic and the Spirit of Capitalism* (1930), including *Roman Agrarian History and Its Significance for Public and Private Law* (1891), *Condition of Farm Labor in Eastern Germany* (1892), and *The Agrarian Sociology of Ancient Civilizations* (1897), in addition to other later essays. These are followed by, among others, Norman Gras's *A History of Agricultural in Europe and America* (1925), Cecil Lewis Gray's *History of Agriculture in the Southern United States* (1933), Joseph Schafer's *The Social History of American Agriculture* (1936), and Percy Bidwell and John Falconer's *History of Agriculture in the Northern United States* (1941). In England, these efforts were matched by early enclosure historians such as Gilbert Slater's *The English Peasantry and the Enclosure of the Common Fields* (1907), J. L. Hammond and Barbara Hammond's *The Village Laborer* (1911), and R. H. Tawney's *The Agrarian Problem in the Sixteenth Century* (1912), among many others.

21. Michel Foucault, *Discipline and Punish: The Birth of the Prison* (New York: Vintage Books, 1979), 205.

22. See, for instance, A. L. Beier, *Masterless Men: The Vagrancy Problem in England, 1560–1640* (London: Methuen, 1985); William Harbutt Dawson, *The Vagrancy Problem: The Case for Measures of Restraint for Tramps, Loafers, and Unemployables: With a Study of Continental Detention Colonies and Labour Houses* (London: P. S. King and Son, 1910).

23. Michel Foucault, *Discipline and Punish: The Birth of the Prison* (New York: Vintage Books, 1979), 138.

24. The statute was passed by parliament (made up at this time of lords that were barons) and enabled the lords to enclose without permission or appeal to the freeholders and villeins.

25. Giorgio Agamben, *Homo Sacer: Sovereign Power and Bare Life*, trans. Daniel Heller-Roazen (Stanford, Calif.: Stanford University Press, 1998), 1.

26. Thirsk, "The Common Fields," 224.

27. For a fascinating and extensive account of enclosure riots and social protest, see Roger B. Manning, *Village Revolts: Social Protest and Popular Disturbances in England, 1509–1640* (Oxford: Clarendon, 1988).

28. Ibid., 228.

29. John Thirsk, "Agricultural Policy: Public Debate and Legislation" from *The Agrarian History of England and Wales*, vol. 5, pt. 2, *1640–1750* (Cambridge: Cambridge University Press, 1985), 318–19, 372.

30. Joan Thirsk, *Tudor Enclosures* (London: Historical Association, 1958), 21n1.

31. John Fletcher Harrison, *The Common People of Great Britain: A History from the Norman Conquest to the Present* (Bloomington: Indiana University Press, 1985).

32. Michael Turner, *English Parliamentary Enclosure* (Kent, England: Wm. Dawson and Sons, 1980), 54. Mark Overton offers a more conservative estimate that the increase was only one-third in yield. See Overton, *Agricultural Revolution in England: The Transformation of the Agrarian Economy 1500–1850* (Cambridge: Cambridge University Press, 1996), 165–66. However, the point is that enclosure advocates justified their position by emphasizing the substantial increase in yields. The self-evident value of greater production—"improvement"—was an essential part of the discourse of enclosure and reflected a new orientation of land management. It was thought that preenclosure "strip farming," in which farmers worked side by side and moved repeatedly between three fields or more, was a wasteful use of time and space. Staying on one area of land was considered more efficient because the farmer could devote the entire time to the planting, maintaining, and harvesting of crops.

33. Michael Turner, *Enclosures in Britain: 1750–1830* (London: Macmillan, 1984), 23.

34. From Sinclair's memoirs reported in E. Halevy, *A History of the English People in the Nineteenth Century*, vol. 1, *England in 1815* (New York: Barnes and Noble, 1961), 230. Quoted in Michael Turner, *English Parliamentary Enclosure: Its Historical Geography and Economic History* (Kent, England: Wm. Dawson and Sons, 1980), 88.

35. Turner, *Enclosures in Britain*, 23.

36. Edward Hasted, *The History and Topographical Survey of the County of Kent*, vol. 9 (London: Wakefield, 1972), 180–81.

37. Ibid., vol. 8, 79.

38. Robert C. Allen, *Enclosure and the Yeoman* (Oxford: Clarendon, 1992), 1.

39. *Encyclopaedia Britannica*, 18 vols. (Cambridge, England: 1797), 1:249.

40. See John Pound, *Poverty and Vagrancy in Tudor England* (London: Longman Group, 1971). "Between one-quarter and one-third of the population were considered to be too poverty-stricken to be assessed for taxation purposes in the period 1523–1527. . . . The wage-earner's position was a tenuous one. Any one who was maintaining himself at one moment might find himself without visible means of support at the next. The rural labourer was better off than his urban counterpart in this respect, for whereas a town-dweller seldom had any alternative means of support, a countryman could often supplement his earnings from the produce of such land as he possessed" (79). "During the period 1500 to 1640, despite a threefold increase in monetary terms, real wages dropped by as much as 50 percent. Those that had to rely on wages alone, and their number increased as the century progressed, invariably found themselves in real difficulties, and their plight was made worse by the fact that there was no guarantee of regular employment" (80).

41. Robert C. Allen, *Enclosures and the Yeoman* (Oxford: Clarendon, 1992), 1.

42. For an example of sustainable relations, see my discussion of the "mead stick" technique of the open-field system. Robert Marzec, *An Ecological and Postcolonial Study of Literature* (New York: Palgrave Macmillan, 2007), 42–51.

43. Michel Foucault, *Security, Territory, Population: Lectures at the Collège de France, 1977–1978* (New York: Picador, 2007).

44. Mitchell M. Dean, for instance, points out that the concept of governmentality includes an increased attention paid toward methods of calculation when it comes to producing forms of knowledge about a population and that increased forms of calculation are directed toward human conduct to "regulate, control, shape and turn to specific ends" (18). "To analyse government is to analyse those practices that try to shape, sculpt, mobilize and work through the choices, desires, aspirations, needs, wants and lifestyles of individuals and groups" (20). See Dean, *Governmentality: Power and Rule in Modern Society* (London: Sage Publications, 2010). See also Majia Holmer Nadesan, *Governmentality, Biopower, and Every Life* (London: Routledge, 2008).

45. Jacques Donzelot, "Pleasure in Work," in *The Foucault Effect: Studies in Governmentality* (Chicago: University of Chicago Press, 1991), 251–80. See also Donzelot's other essay in the same volume, "The Mobilization of Society," 169–80.

46. The amount of scholarship that mentions, and in some cases thematizes, the role of enclosure in the development of the modern nation is growing. A selection would include Carlos J. Alonso, *The Burden of Modernity: The Rhetoric of Cultural Discourse in Spanish America* (New York: Oxford University Press, 1998); Vandana Shiva, Asfar H. Jafiri, Gitanjali Bedi, and Radha Holla-Bhar, *The Enclosure and Recovery of the Commons* (New Delhi, India: Research Foundation for Science, Technology, and Ecology, 1997); Antonio Gomez Sal, Fernando Gonzalez Bernaldez, Francesco di Castri, and B. L. Turner, eds., *Global Land Use Change: A Perspective from the Columbian Encounter* (Madrid, Spain: Consejo Superior de Investigacions Cientificas, 1995); Daniel W. Bromley, "The Enclosure Movement Revisited: The South African Commons," *Journal of Economic Issues* 28, no. 2 (June 1994): 357–65; Tor A. Benjaminsen, *Securing Land Rights in Africa* (London: Frank Cass, 2003); Bina Agarwal, *A Field of One's Own: Gender and Land Rights in South Asia* (Cambridge: Cambridge University Press, 1994); Wayne Lee, ed., *Empires and Indigenes: Intercultural Alliance, Imperial Expansion, and Warfare in the Early Modern World* (New York: New York University Press, 2011).

47. UN Development Programme, *Human Development Report 2007/2008. Fighting Climate Change: Human Solidarity in a Divided World* (New York: UN Development Programme, 2007).

48. Miguel Carter, "The Landless Rural Workers Movement and Democracy in Brazil," *Latin American Research Review* 45, no. 4 (2010): 186–217, 189.

49. Ibid., 191–92.

50. Daniel do Nascimento e Silva, "Identities Forged in Pain and Violence: Nordeste's Writing" (paper presented at the 2012 Congress on Latin American Studies, Toronto, October 6–9, 2010), http://lasa.international.pitt.edu/members/congress-papers. See also Miguel Carter, "The Landless Rural Workers' Movement (MST) and Democracy in Brazil" (working paper, University of Oxford/Center of Brazilian Studies, 2005), http://www.brazil.ox.ac.uk.

51. Monica Weinberg, "Madrassas MST: Like the Muslim Boarding Schools, Landless Teach Hatred and Incite Revolution. The Infidel Is All of Us," *Veja*, September 8, 2004.

52. Arturo Alvarado, "The Militarization of Internal Security and Its Consequences for Democracy: A Comparison between Brazil, Mexico, and Colombia," *APSA 2010 Annual Meeting Paper*, papers.ssrn.com.

53. Carter, "The Landless Rural Workers' Movement," 186. See also Sue Branford and Jan Rocha, *Cutting the Wire: The Story of the Landless Movement in Brazil* (London: Latin American Bureau, 2002), 203.

54. Peter Kornbluh, "Brazil Marks 40th Anniversary of Military Coup: Declassified Documents Shed Light on U.S. Role," *National Security Archive*, http://www.gwu.edu.

55. Angus Wright and Wendy Wolford, *To Inherit the Earth: The Landless Movement and the Struggle for a New Brazil* (Oakland, Calif.: Food First Books, 2003), 130.

56. See M. Kato, M. L. Rocha, A. B. Carvalho, M. E. Chaves, M. C. Raña, and F. C. Oliveira, "Occupational Exposure to Neurotoxicants: Preliminary Survey in Five Industries of the Camaçari Petrochemical Complex, Brazil," *Environmental Research* 61, no. 1 (April 1993): 133–39.

57. For more on recent activity in Brazilian multinational operations, see Luiz Caseiro, "Destination and Strategy of Brazilian Multinationals," *Economics, Management and Finance Markets* 6, no. 1 (March 2011): 207–38.

58. Sue Branford, "From Nothing to Nowhere—The Transamazonian Highway," *New Internationalist Magazine*, October 1, 1980.

59. "Logging in the Amazon," *Greenpeace*, November 21, 2005, http://www.greenpeace.org.

60. Wright and Wolford, *To Inherit the Earth*, 187.

61. Raymond Colitt, "Brazil Amazon Forest to be Privately Managed," *Reuters*, October 12, 2010, http://www.reuters.com.

62. For a discussion of educational extension services, see Sandra Maria Gadelha de Carvalho and José Ernandi Mendes, "The University and the Landless Movement in Brazil: The Experience of Collective Knowledge Construction through Educational Projects in Rural Areas," in *Social Movements in the Global South: Dispossession, Development and Resistance*, ed. Sara C. Motta and Alf Gunvald Nilsen (London: Palgrave Macmillan, 2011), 131–49.

63. See Ray Walser, "U.S.-Brazil Summit: Time for a Gold Policy Vision," *The Heritage Foundation*, April 6, 2012, http://www.heritage.org.

64. See Nestor Bailly, "Brazil Military Ties with U.S. Increasing," *Rio Times*, December 14, 2010, http://riotimesonline.com. See also the FAS (Federation of American Scientists) website for more specifics on Brazil and its military history: http://www.fas.org.

65. This is Ray Walser's phrase. See Walser, "U.S.-Brazil Summit."

66. One upward estimate put Brazil's offshore fields at eight billion barrels. See Carlos Caminada and Jeb Blount, "Petrobras' Tupi Oil Field May Hold 8 Billion Barrels (Update6)," *Bloomberg*, November 8, 2007, http://www.bloomberg.com. Though the Jupiter field is larger (and is also important because it contains deposits of gas), the amount of oil there is thought to be roughly the same. The Heritage Foundation's Ray Walser puts the total of the Tupi and the Jupiter fields at one hundred billion barrels. I have not been able to discover the source of this figure. For specifics on rankings of world oil fields, see Rich Minerals Corporation, "Three Super-Giant Fields Discovered Offshore Brazil," July 23, 2008, http://www.richminerals.ca/m1.html.

67. David Fessler, "The Oil Market's 'New World Order,'" *Investment U*, February 18, 2011, http://www.investmentu.com.

68. Gary Duffy, "'Huge' Gas Field Found off Brazil," *BBC News*, January 22, 2008, http://news.bbc.co.uk.

69. See "U.S.-Brazil Defense Cooperation Agreement," *Just the Facts*, April 13, 2010, http://Justf.org/blog/2010/04/13/us-brazil-defense-cooperation-agreement. See also Pincas Jawetz, "Expanding Alliances in the 21st Century: The U.S. and Brazil Unite to Address Matters of National Security," *SustainabiliTank*, August 21, 2010, http://www.sustainabilitank.info.

70. See Jim Garamone, "Dempsey Looks to Expand Military Ties with Brazil," *U.S. Department of Defense*, March 29, 2012, http://www.defense.gov.

71. See Jackie Calmes and Alexei Barrionuevo, "Amid Crises, Obama Lands in South America," *New York Times*, March 19, 2011, http://www.nytimes.com.

72. See Kent Klein, "Obama to Host Brazil's Rousseff Monday," *Voice of America*, April 7, 2012, http://www.voanews.com.

73. Though publications on security have a longer history (i.e., the journal *International Security* began publication in 1976, *Terrorism* in 1977, *Intelligence and National Security* in 1986, *Global Change, Peace and Security* in 1989, *Science and Global Security* in 1989), a brief survey reveals that academic and nongovernmental organization–sponsored journals that connect security to energy, food, water, and other resources is relatively recent: *Agriculture and Food Security* (2012), *Food and Energy Security* (2012), *Food Security* (2009), *Journal of Energy Security* (2008).

74. See Raphael D. Sagarin and Terence Taylor, *Natural Security: A Darwinian Approach to a Dangerous World* (Berkeley: University of California Press, 2008).

75. Gawdat Bahgat, *Energy Security: An Interdisciplinary Approach* (London: Wiley, 2011), 1.

76. Jan H. Kalicki and David L. Goldwyn, eds., *Energy and Security: Toward a New Foreign Policy Strategy* (Baltimore: Johns Hopkins University Press, 2005), esp. 1–17.

77. Amanda Griscom Little, "Who's Afraid of the Big, Bad Woolsey?," *Grist* 7 (2008), http://www.grist.org.

78. See the official Green Patriot website: http://www.greenpatriot.us.

79. Woolsey's statement is found on the main page of the Set America Free Coalition website: http://www.setamericafree.org.

80. Frank J. Gaffney, *War Footing: 10 Steps America Must Take to Prevail in the War for the Free World* (Annapolis, Md.: Naval Institute Press, 2006), 53.

81. Information on the new "EcoPartnerships" can be found on the following website: http://www.ecopartnerships.gov.

82. Joseph Stiglitz and Andrew Charlton, *Fair Trade for All* (Oxford: Oxford University Press, 2005), 57.

83. Kym Anderson and Will Martin, "Agriculture, Trade Reform, and the Doha Agenda," in *Agricultural Trade Reform and the Doha Development Agenda*, ed. Kym Anderson and Will Martin (New York: Palgrave Macmillan, 2006), 11.

84. Chad E. Hart and John C. Beghin, "Rethinking Agricultural Domestic Support under the World Trade Organization," in *Agricultural Trade Reform and the Doha Development Agenda*, ed. Kym Anderson and Will Martin (New York: Palgrave Macmillan, 2006), 239.

85. The enclosure movement is not as foregrounded in U.S. history as it is in English history. Nevertheless, enclosure was transplanted to the United States from England in both theory and practice. See, for instance, William Cronon, *Changes in the Land: Indians, Colonists, and the Ecology of New England* (New York: Hill and Wang, 2003).

86. Jan H. Kalicki and David L. Goldwyn, "Introduction: The Need to Integrate Energy and Foreign Policy," in *Energy and Security: Toward a New Foreign Policy Strategy*, ed. Jan Kalicki and David L. Goldwyn (Washington, D.C.: Woodrow Wilson Center Press, 2005), 5. Kalicki is public policy scholar at the Woodrow Wilson International Center for Scholars and counselor for international strategy at Chevron Corporation. Goldwyn is president of Goldwyn International Strategies, was U.S. assistant secretary of energy for international affairs from 2000 to 2001, and was national security deputy to former U.S. Ambassador to the United Nations Bill Richardson.

87. Ibid., 2.

88. Andréa Zhouri, "'Adverse Forces' in the Brazilian Amazon: Developmentalism versus Environmentalism and Indigenous Rights," *Journal of Environment and Development* 19, no. 3 (2010): 252–73, 255.

89. Ibid., 256.

90. Robert Zubrin, *Energy Victory: Winning the War on Terror by Breaking Free of Oil* (Amherst, N.Y.: Prometheus Books, 2007), 147.

91. Francisco Vidal Luna and Herbert S. Klein, *Brazil since 1980* (Cambridge: Cambridge University Press, 2006), 49.

92. Zubrin, *Energy Victory*, 170.

93. These policies of energy security differ from former Cold War policies of communist containment, which operated through the establishment of state-to-state power relations (politico-economic agreements between the United States and other state authorities), whereas power relations in the discourse of energy security function by anchoring control directly on the land—that is, on the correct use of the land and the most efficient means to reap the total benefits the land has to offer.

94. Sérgio Sauer, "The World Bank's Market-Based Land Reform in Brazil," in *Promised Land: Competing Visions of Agrarian Reform*, ed. Peter Rosset, Raj Patel, and Michael Courville (Oakland, Calif.: Food First Books, 2006), 178.

95. Wright and Wolford, *To Inherit the Earth*, xv.

96. Patrick J. McDonnell, "Human Cost of Brazil's Biofuels Boom," *Los Angeles Times*, June 16, 2008, http://www.latimes.com.

97. Amnesty International, *Amnesty International Annual Report, New York, 2008*, http://www.refworld.org/docid/483e277d40.html.

98. Huw Beynon and José R. Ramalho, "Democracy and the Organization of Class Struggle in Brazil," in *Working Class Global Realities: Socialist Register 2001*, ed. Leo Panitch and Colin Leys (London: Merlin, 2000), 217.

99. Wright and Wolford, *To Inherit the Earth*, 114.

100. J. H. Baker, *An Introduction to English Legal History*, 4th ed. (Oxford: Oxford University Press, 2002), 38.

101. A. H. Manchester, *Modern Legal History of England and Wales, 1750–1950* (London: Butterworth-Heinemann, 1980), 128.

102. Thomas Arnold Herbert, *The History of the Law of Prescription in England* (London: C. J. Clay and Sons, 1891), 66.

103. Ibid.

104. Ibid., 67.

105. Ibid., 69.

106. Andrea C. Loux, "The Persistence of the Ancient Regime: Custom, Utility, and the Common Law in the Nineteenth Century," *Cornell Law Review* 79, no. 183 (1993): 183–218, 191.

107. See Coke, *Court of Common Please*, 6, fo. 59.

108. E. P. Thompson, *Customs in Common* (New York: Penguin, 1991), 133–34.

109. Loux, "The Persistence of the Ancient Regime," 192.

110. Ibid., 191.

111. Ibid., 191–92.

112. Andrew Thrush and John P. Ferris, *The History of Parliament: The House of Commons 1604–1629* (Cambridge: Cambridge University Press, 2010).

113. Harlod J. Berman, *Law and Revolution II: The Impact of the Protestant Reformations on the Western Legal Tradition* (Cambridge, Mass.: Harvard University Press, 2003), 333.

114. Ibid., 334.

115. John G. Bellamy, *The Tudor Law of Treason: An Introduction* (London: Taylor and Francis, 1979), 80.

116. J. A. Sharpe, "Social Strain and Social Dislocation, 1585–1603," in *The Reign of Elizabeth I: Court and Culture in the Last Decade*, ed. John Guy (Cambridge: Cambridge University Press, 1995), 192–211.

117. Loux, "The Persistence of the Ancient Regime," 213.

118. Karl Marx and Frederick Engels, *The German Ideology Part One, with Selections from Parts Two and Three, and Supplementary Texts* (New York: International Publishers, 1970), 46. See also Karl Marx, "Precapitalist Economic Formations," in *Grundrissse: Foundations of the Critique of Political Economy*, trans. Martin Nicolaus (New York: Penguin, 1993), 471–514.

119. Loux, "The Persistence of the Ancient Regime," 213.

120. See Robert Marzec, *An Ecological and Postcolonial Study of Literature*, 12–13.

121. Daniel Greenberg and Frederick Stroud, *Stroud's Dictionary of Words and Phrases*, 7th ed. (London: Sweet and Maxwell, 2006), 387.

122. Ibid., 388.

123. Benjamin W. Pope, *Legal Definitions: A Collection of Words and Phrases as Applied and Defined by the Courts, Lexicographers and Authors of Books on Legal Subjects*, vol. 1 (Chicago: Callaghan, 1919), 760.

124. Henry Campbell Black, *A Law Dictionary*, 2nd ed. (St. Paul, Minn.: West, 1910), 625.

125. Greenberg and Stroud, *Stroud's Dictionary*, 388.

126. Ernesto Laclau, *Emancipation(s)* (London: Verso, 1996), 15.

127. Brenda Baletti, Tamara M. Johnson, and Wendy Wolford, "Late Mobilization: Transnational Peasant Networks and Grassroots Organizing in Brazil and South Africa," *Journal of Agrarian Change* 8, nos. 2–3. (April and July 2008): 291; 290–314.

128. Mônica Dias Martins, "Learning to Participate: The MST Experience in Brazil," in *Promised Land: Competing Visions of Agrarian Reform*, ed. Peter Rosset, Raj Patel, and Michael Courville (Oakland, Calif.: Food First Books, 2006), 269, 265–76. See also in the same volume Beatriz Heredia, Leonilde Medeiros, Moacir Palmeira, Rosângela Cintrão, and Sérgio Pereira Leite, "Regional Impacts of Land Reform in Brazil," 277–300.

129. Cliff Welch, "Movement Histories: A Preliminary Historiography of the Brazil's Landless Laborers' Movement (MST)," *Latin American Research Review* 41,

no. 1 (2006): 198–210. These are 2006 figures. Wright and Wolford report 2002 figures as "approximately 350,000 families in three thousand settlements" (75).

130. Wright and Wolford, *To Inherit the Earth*, 60.

131. Martins, "Learning to Participate," 273.

132. Jeffrey W. Rubin, "From Che to Marcos," *Dissent* 49, no. 3 (Summer 2002): 45–46.

133. "What Is Happening to Agrobiodiversity?," *Food and Agricultural Organization of the United Nations*, http://www.fao.org.

134. Ibid.

135. Ibid.

136. Vandana Shiva, *Monocultures of the Mind: Perspectives on Biodiversity and Biotechnology* (London: Zed Books, 1993), 49.

137. Peter J. Jacques and Jessica R. Jacques, "Monocropping Cultures into Ruin: The Loss of Food Varieties and Cultural Diversity," *Sustainability* 4, no. 11 (2012): 2970–97.

138. Shiva, *Biopiracy*, 48.

139. Maria Luisa Mendona, "Monocropping for Agrofuels: The Case of Brazil," *Development* 54, no. 1 (2011): 98–103, 98.

140. Ibid., 99.

141. See Donald Pisani, *From the Family Farm to Agribusiness: The Irrigation Crusade in California and the West, 1950–1931* (Berkeley: University of California Press, 1984).

142. Maria Luisa Mendonça, Isidoro Revers, Marluce Melo, and Plácido Júnior, "*Impactos da produção de cana no Cerrado e Amazônia* [Impacts of the production of sugarcane in the Cerrado and the Amazon]," Pastoral Land Commission, December 2008.

143. Wright and Wolford, *To Inherit the Earth*, 150.

144. See Vandana Shiva, *The Violence of the Green Revolution: Third World Agriculture, Ecology, and Politics* (London: Zed Books, 1991); Harry Cleaver, "The Contradictions of the Green Revolution," *American Economic Review* 62, no. 2 (May 1972): 177–86.

145. Wright and Wolford, *To Inherit the Earth*, 94.

146. Michael Hardt and Antonio Negri, *Multitude: War and Democracy in the Age of Empire* (New York: Penguin, 2004), 113.

147. Wright and Wolford, *To Inherit the Earth*, 93, 95.

148. Saturnino M. Borras Jr., Marc Edelman, and Cristóbal Kay, "Transnational Agrarian Movements: Origins and Politics, Campaigns and Impact," *Journal of Agrarian Change* 8, nos. 2–3 (April/July 2008): 169–204, 183.

149. João Pedro Stédile, "The Class Struggles in Brazil: The Perspective of the MST," in *Global Flashpoints: Reactions to Imperialism and Neoliberalism, Socialist Register 2008*, ed. Leo Panitch and Colin Leys (London: Merlin, 2008), 193–216, 195.

150. "Global Small-Scale Farmers' Movement Developing New Trade Regimes," *News & Views* 28, no. 97 (Spring/Summer 2005): 2. See also Saturnino M. Borras Jr., "La Via Campesina and Its Global Campaign for Agrarian Reform," *Journal of Agrarian Change* 8, nos. 2–3 (April 2008): 258–89.

151. Gilles Deleuze and Félix Guattari, *A Thousand Plateaus: Capitalism and Schizophrenia* (Minneapolis: University of Minnesota Press, 1987), 353.

152. Angus Wright and Wendy Wolford, *To Inherit the Earth: The Landless Movement and the Struggle for a New Brazil* (Oakland, Calif.: Food First Books, 2003), 95.

153. Vandana Shiva, *Biopiracy: The Plunder of Nature and Knowledge* (Boston: South End, 1997).

154. Wright and Wolford, *To Inherit the Earth*, 98.

155. Jacques Rancière, *Disagreement: Politics and Philosophy* (Minneapolis: University of Minnesota Press, 1999), 8.

156. Write and Wolford, *To Inherit the Earth*, 84.

157. Ibid., 24.

158. Ibid., 107.

159. Rancière, *Disagreement*, 10.

160. This is Rancière's phrase: "Politics exists through the fact of a magnitude that escapes ordinary measurement, this part of no part that is nothing and everything. This paradoxical magnitude has already pulled the plug on market measures." Rancière, *Disagreement*, 15.

161. Deleuze and Guattari, *A Thousand Plateaus*, 382.

162. Maros Ivanic and Will Martin, "Implications of Higher Global Food Prices for Poverty in Low-Income Countries," *Policy Research Working Paper 4594*, the World Bank Development Research Group Trade Team, April 2008, http://econ.worldbank.org.

3. Genealogies of Military Environmentality I

1. Atomic Heritage Foundation, "Little Boy and Fat Man," http://www.atomicheritage.org/history/little-boy-and-fat-man.

2. O. R. Frisch, "Eyewitness Account of 'Trinity' Test, July 1945," in *The American Atom: A Documentary History of Nuclear Politics from the Discovery of Fission to the Present*, ed. Philip L. Cantelon, Richard G. Hewlett, and Robert C. Williams (Philadelphia: University of Pennsylvania Press, 1984), 50–51.

3. Craig Nelson, *The Age of Radiance: The Epic Rise and Dramatic Fall of the Atomic Era* (New York: Scribner, 2014), 100.

4. Ibid., 100.

5. Dipesh Chakrabarty, "Postcolonial Studies and the Challenge of Climate Change," *New Literary History* 43, no. 1 (Winter 2012): 1–18, 2.

6. For a detailed account of this concept and its history, see W. Poundstone, *Prisoner's Dilemma* (New York: Doubleday, 1992).

7. Philip Mirowski, *Machine Dreams: Economics Becomes a Cyborg Science* (Cambridge: Cambridge University Press, 2002), 261.

8. The report was written in late 1944 and appeared as Memorandum No. 8 of the Fire Control Research Office at Princeton. It was distributed as a working paper through the American military personnel and declassified in 1948. It was later published as a Project RAND memorandum (*RM-913*, "Aerial Bombing Tactics: General Considerations [A World War II Study]"), dated September 2, 1952. Robert Leonard, *Von Neumann, Morgenstern, and the Creation of Game Theory: From Chess to Social Science, 1900–1960* (Cambridge: Cambridge University Press, 2010), 281.

9. For a discussion of these connections from the standpoint of an economist, see Mirowski's *Machine Dreams*, especially chap. 2, "Some Cyborg Genealogies; or, How the Demon Got Its Bots," 26–92.

10. T. M. Within, "Decision-Making and the Theory of Organization," sponsored by the Econometric Society and the Institute of Mathematical Statistics, September 6, 1951, copy in box 22, OMPD. Quoted in Mirowski, *Machine Dreams*, 355–56.

11. Ibid., 356.

12. Manuel De Landa, *War in the Age of Intelligent Machines* (New York: Zone Books, 1991), 155.

13. Merrill Flood, "What Future Is There for Intelligent Machines?," *General Systems* 8: 219–26.

14. Michel Foucault, *Discipline and Punish: The Birth of the Prison*, trans. Alan Sheridan (New York: Pantheon Books, 1977).

15. See Manual De Landa's lecture, "The Biology of Cities 1," from a seminar given at the USC School of Architecture in 2011 titled "Theories of Self-Organization and Dynamics of Cities," http://www.egs.edu.

16. Arthur Iberall, "A Physics for the Study of Civilizations," in E. Yates, ed., *Self-Organizing Systems: The Emergence of Order* (New York: Plenum Press, 1987), 531–33.

17. Christopher W. Kuzawa and Zaneta M. Thayer, "Timescales of Human Adaptation: The Role of Epigenetic Processes," *Epigenomics* 3, no. 2 (April 2011): 221–34, http://www.futuremedicine.com.

18. Andrew Angel, Jie Song, Caroline Dean, and Martin Howard, "A Polycomb-Based Switch Underlying Quantitative Epigenetic Memory," *Nature* 476 (2011): 105–8, doi:10.1038/nature10241.

19. Barton C. Hacker, *American Military Technology: The Life Story of a Technology* (Baltimore: Johns Hopkins University Press, 2007), 10–19.

20. De Landa, *War in the Age of Intelligent Machines*, 4.

21. Ibid., 12.

22. "Jäger History and Facts," http://www.webcitation.org.

23. As a concept, "singularity" has a theoretical tradition of signifying an event that cannot be repeated and, as such, has nothing to do with a uniform system of mass production. De Landa will use the term in this fashion but also to connote something like a sudden takeoff of a distinct development or phenomenon that defines military activity in the post-Enlightenment era. The shift to the generalized logistical economy of the "projectile wall" organization is named as a new singularity in the sense of a radical departure from previous organizations. However, this particular shift—one that marks what Heidegger would have called the "genuine event of history" (in the sense of an event that starts a new historical line of production and development)—brings about a radical change in relation to singularities. Previous organizational procedures—self-organizing events—still had what we might call a healthy or open relation to other singularities so that future connections and effects were possible. The military essence of the projectile wall, however, precludes—or attempts to preclude—future singularities from forming. It is what we might call a "final" or "completed" singularity, and its essential characteristic is its maintaining of a generalized economy that specifically targets other singularities so as to destroy them (the exterior-enemy dynamic analyzed in the previous chapter). It is therefore necessary to make a distinction between singularities that maintain an open relationship with other singularities and singularities that develop into universalized, higher-order motors of domination.

24. De Landa, *War in the Age of Intelligent Machines*, 31.

25. See Merritt Roe Smith, "Army Ordnance and the 'American System' of Manufacturing, 1815–1861," in *Military Enterprise and Technological Change*, ed. M. R. Smith (Cambridge, Mass.: MIT Press, 1987), 39–86.

26. De Landa, *War in the Age of Intelligent Machines*, 32.

27. "*Ta mathemata* means for the Greeks that which man knows in advance in his observation of whatever is and in his intercourse with things: the corporeality of bodies, the vegetable character of plants, the animality of animals, the humanness of man. Alongside these, belonging also to that which is already-known, i.e., the mathematical, are numbers. If we come upon three apples on the table, we recognize that there are three of them. But that number three, threeness, we already know. This means that number is something mathematical. Only because numbers represent, as it were, the most striking of always-already-knowns and thus offer the most familiar instance of the mathematical, is 'mathematical' promptly reserved as a name for the numerical. In no way, however, is the essence of the mathematical defined by numberness." Martin Heidegger, "The Age of the World Picture," in *The Question Concerning Technology and Other Essays* (New York: HarperCollins, 1982), 118–19.

28. Ibid., 131–32.

29. David F. Noble, "Command Performance: A Perspective of Military Enterprise and Technological Change," in Smith, *Military Enterprise*, 332–33.

30. Ibid., 333.

31. De Landa, *War in the Age of Intelligent Machines*, 34.

32. Ibid., 36.

33. Ibid., 41.

34. Jamie Gumbrecht, "Rediscovering WWII's Female 'Computers,'" *CNN*, February 8, 2011, http://www.cnn.com.

35. Noah Wardrip-Fruin and Nick Montfort, eds., *The New Media Reader* (Cambridge, Mass.: MIT Press, 2003), 35.

36. Vannevar Bush, "As We May Think," *The Atlantic*, July 1945.

37. John W. Dower, *Cultures of War: Pearl Harbor, Hiroshima, 9/11, Iraq* (New York: W. W. Norton, 2010), 427.

38. Thomas P. Coakley, *C3i: Issues of Command and Control* (Washington, D.C.: National Defense University, 1991), xvi–xvii.

39. Ibid., xvii.

40. "Communications, Command, and Control (C3)," http://www.global security.org.

41. "Chapter 20: Command, Control, and Communication," in *Fundamentals of Naval Weapons Systems* (United States Naval Academy, n.d.), http://fas.org/man/dod-101/navy/docs/fun/part20.htm.

42. United States Naval Academy, *Fundamentals of Naval Weapons Systems*, http://fas.org/man/dod-101/navy/docs/index.html.

43. Claude Elwood Shannon, "A Mathematical Theory of Communication," *Bell System Technical Journal* 27 (July/October 1984): 379–423, 623–56, http://cm.bell-labs.com. See also K. Miller, *Communication Theories: Perspectives, Processes, and Contexts* (New York: McGraw-Hill, 2005).

44. De Landa, *War in the Age of Intelligent Machines*, 72.

45. Mark Greenia, "ARPA: The Advanced Research Projects Agency: The Financial Backbone of U.S. Computer Research," in *The History of Computing* (Lexicon Services, 2003), http://www.computermuseum.li/Testpage/99HISTORYCD-ARPA-History.HTM.

46. "What Keeps DARPA Leadership up at Night," February 29, 2012, http://www.darpa.mil.

47. Ibid.

48. This is in part an extension of Paul Vilirio's observation that the airplane camera photography accumulated during World War I was the optical: "the supply of images became the equivalent of an ammunition supply" (1). Paul Virilio, *War and Cinema: The Logistics of Perception* (London: Verso, 1989).

49. Ibid., 2.

50. Ibid., 3.

51. Ibid., 15.

52. Ibid., 15.

53. De Landa, *War in the Age of Intelligent Machines*, 22.

54. Ibid., 24.

55. Ibid., 69.

56. Ibid., 69.

57. Ibid., 155.

58. See Gilles Deleuze and Félix Guattari, *A Thousand Plateaus: Capitalism and Schizophrenia* (Minneapolis: University of Minnesota Press, 1987), 467.

59. The relation between a militarized existence and the rise of computers is, of course, much more complex than I have the time to address here. De Landa himself, at the end of *War in the Age of Intelligent Machines*, in a rather hopeful and somewhat naïve gesture, explores the very different potentials that can be put into practice by humans interacting with computers. The development of the personal computer introduced new directions that exceeded the seemingly totalizing mandates of mechanized thought processes. He cites the work of anarchist computer hackers such as Steve Jobs, who worked to open new platforms of communal, nonmilitarized connections between humans vis-à-vis the new "windows" of information flow on the computer screen—windows that open new lines of connection between humans working in different areas and fields, which in turn opened radically decentralized interactions. At such moments, argues De Landa, the military dream of a tightly controlled chain of command becomes so thoroughly disrupted as to be impossible. I say that this is somewhat naïve because at this moment in his analysis, De Landa oddly becomes less cautious and separates the being of information-processing technology from the military, as if the very idea of information processing did not, as he argues throughout the book, come into existence precisely to solve military matters: "Information-processing technology is a key branch of the machinic phylum and, in a sense, it has been made hostage by military institutions. One has only to think about the NSA's commitment to stay five years ahead of the state of the art in computer design to realize that the cutting edge of digital technology is being held hostage by paramilitary organizations. Hackers and visionary scientists have opened small escape routes for the phylum, developing interactivity in order to put computer power in everybody's hands" (229–30). This argument for establishing a mode of interaction with computer technology outside of the military's influence depends on us not thinking about the specific essence of both "interactivity" and the "processing" of "information." Does the military not also depend on a decentered interactivity, as we've seen from the examples of intelligence gathering and the need to shift from self-destructive top-down structures to more open motor structures?

4. Genealogies of Military Environmentality II

1. The Center for Naval Analysis Corporation, *National Security and the Threat of Climate Change*, 2007, 10, https://www.cna.org.

2. The IPCC's "Assessment Reports" are released at different stages throughout the year. The approval of the last "synthesis" stage of the report and the release of the full report was September 29, 2001. See United Nations Framework Convention on Climate Change online information given in these reports: "IPCC Assessment Process" and "Cooperation with the Intergovernmental Panel on Climate Change," http://unfccc.int/science/workstreams/cooperation_with_the _ipcc/items/1077.php.

3. "Biopolitical settlement" is Donald Pease's phrase for the status of human existence in the wake of the transformation of the myth of "virgin land" into the "Homeland" by the Bush administration. See Pease, *The New American Exceptionalism* (Minneapolis: University of Minnesota Press, 2009).

4. Harlan K. Ullman and James P. Wade, *Shock and Awe: Achieving Rapid Dominance* (Washington, D.C.: National Defense University, 1996), xxiv; Giorgio Agamben, *State of Exception,* trans. Kevin Attell (Chicago: University of Chicago Press, 2005), 1.

5. This statement is from the Center for Naval Analysis Corporation, *National Security and the Threat of Climate Change*, 18.

6. William V. Spanos, *American Exceptionalism in the Age of Globalization: The Specter of Vietnam* (Albany: State University of New York Press, 2008), 10.

7. Ibid., 22.

8. Pease, *The New American Exceptionalism*, 24.

9. Sacvan Bercovitch, *The American Jeremiad* (Madison: University of Wisconsin Press, 1978), 23.

10. Michel Foucault, *The Birth of Biopolitics: Lectures at the Collége de France, 1978–1979* (New York: Palgrave Macmillan, 2008), 19.

11. Agamben, *State of Exception*, 5. Emphasis added.

12. Ibid., 5.

13. Ibid., 22.

14. Ibid., 5–6. Emphasis added.

15. Barry Commoner, *Science and Survival* (New York: Viking, 1967), 25.

16. Foucault, *The Birth of Biopolitics*, 19.

17. Sharon Burke and Christine Parthemore, "Climate Change War Game: Major Findings and Background" (working paper, Center for a New American Security, Washington, D.C., June 2009), 8.

18. Though, as we will see, even this iron law can be pushed beyond its limit to open a new threshold of military sovereignty. Argues retired Marine Corps general Anthony Zinni, former commander of U.S. forces in the Middle East, "We

will pay to reduce greenhouse gas emissions today, and we'll have to take an economic hit of some kind. Or, we will pay the price later in military terms. And that will involve human lives. There will be a human toll." Quoted in John M. Broder, "Climate Change Seen as Threat to U.S. Security," *New York Times*, August 8, 2009, http://www.nytimes.com/2009/08/09/science/earth/09climate.html?page wanted=all&_r=0.

19. Peak oil production is in part an economic concept; production will decline as other sources become more available. There are various models for and thus some disagreement about establishing peak oil production and decline. American oil production actually peaked in 1971, a date that was successfully predicted by Marion King Hubbert in 1956. See Adam R. Brandt, "Testing Hubbert," *Energy Policy* 35, no. 5 (May 2007): 3074–88. R. W. Allmendinger at Cornell University places peak oil production between 2020 and 2040. See Allemdinger, "Peak Oil?," in *Energy Studies in the College of Engineering* (Cornell University, 2007), http://www.geo.cornell.edu/eas/energy/the_challenges/peak_oil.html. Some argue that peak oil production has already occurred. See Kjell Aleklett, Mikael Höök, Kristofer Jakobsson, Michael Lardelli, Simon Snowden, and Bengt Söderbergh, "The Peak of the Oil Age," *Energy Policy* 38, no. 2 (November 9, 2009): 1398–1414. Others argue that we are on the cusp. See Nick A. Owen, Oliver R. Inderwildi, and David A. King, "The Status of Conventional World Oil Reserves—Hype or Cause for Concern?," *Energy Policy* 38, no. 8 (2010): 4743. It is also important to point out that "peak oil production" is not the same as the decline of crude oil available on the planet for extraction or the decline of reserves. Most of the planet's crude oil fields have been discovered. However, past oil fields are now classified as "cheap oil" because their extraction was relatively simple compared to more recently discovered forms of oil, which required increasingly expensive and time-consuming methods of extraction and refinement.

20. "A warmer climate will be bad news for global agriculture, with regional winners and losers, says Andrew Challinor of the University of Leeds, U.K. Agronomists are busy designing new varieties of staple crops like rice and wheat able to survive more frequent heatwaves and droughts, and organisations like the Consultative Group on International Agricultural Research are helping farmers find out what works for them. Despite such efforts, crops will fail more often, probably leading to food-price spikes." Michael Marshall, "Earth in Balmy 2080," *NewScientist*, December 5, 2011.

21. Deleuze and Guattari define capitalism as a system that constantly "postpones its own limit"; Gilles Deleuze and Félix Guattari, *A Thousand Plateaus: Capitalism and Schizophrenia* (Minneapolis: University of Minnesota Press, 1987): 436–37.

22. Burke and Parthemore, "Climate Change War Game," 15.

23. Douglas V. Johnson II, comp., "Global Climate Change: National Security Implications," U.S. Army War College and Triangle Institute for Security Studies, May 1, 2007, http://www.strategicstudiesinstitute.army.mil.

24. Carolyn Pumphrey, ed., *Global Climate Change: National Security Implications* (Carlisle, Pa.: Strategic Studies Institute, Army War College, 2008), http://www.strategicstudiesinstitute.army.mil.

25. This catastrophic language was in fact transferred over to the National Security Act of 2010. However, the act placed more emphasis on developing sustainable energy solutions within an economic model that would help to rebuild the American economy. See *National Security Strategy 2010* (Washington, D.C., 2010), 9, https://www.whitehouse.gov/sites/default/files/rss_viewer/national_security_strategy.pdf.

26. The Center for Naval Analysis Corporation, *National Security and the Threat of Climate Change*, 7.

27. Ibid., 11.

28. Ibid., 12.

29. ibid., 10.

30. Scripps Institution of Oceanography, "New Center Addresses National Security Implications of Environmental Change," http://www.sio.ucsd.edu/Announcements/Harnish_CENS.

31. Ibid.

32. Ibid.

33. Ibid.

34. I take this phrase from Slavoj Žižek's early work on ideological dynamics. See "Introduction: The Specter of Ideology," in *Mapping Ideology* (New York: Verso, 1994), 4.

35. Agamben, *State of Exception*, 1.

36. Ibid., 39.

37. At the time of this writing, the *Wall Street Journal* had just published a screed on the "truth" about the global warming hoax, "No Need to Panic about Global Warming," which argued that there was "no compelling scientific argument for drastic action to 'decarbonize' the world's economy." *Wall Street Journal*, January 26, 2012, http://www.wsj.com.

38. The Center for Naval Analysis Corporation, *National Security and the Threat of Climate Change*, 10.

39. In using *(a)political* rather than simply *apolitical*, I mean to mark a difference between the two. *Apolitical* tends to be understood in terms of its connection to a neutral position, despite the fact that, as poststructuralism has repeatedly shown, there is no true position of neutrality; taking up a neutral position in a set of clearly political positions is itself a political act. Apolitical thus has the baggage, within the history of poststructuralism, of needing to be unveiled as actually being a political ideology masking its presence. Instead, by *(a)political*, I mean to designate more clearly the idea of cutting through limiting political agendas in order to take a political action that justifies itself on the fantasy of an empirical

necessity presenting itself as the only choice. In this sense, the action is political but comprehended as not political (not a matter of partisan politics). The point of making such a distinction is to show the proximity with which military (a)politicality shares contemporary theoretical explorations of political resistance: in the forms of Žižek's conceptualizations of the Lacanian "Real" or Rancière's conceptualization of a genuine political act. For a discussion of the apolitical as the most seductive form of ideology, see Žižek, "Introduction: The Spectre of Ideology," 1–33.

40. The "Real" is that which disturbs and disrupts the symbolic order of the fantasy. It is abject in relation to the fantasy, and that abjectness calls the logic of the fantasy into question. In this sense, it appears, at first, as if the military action that addresses environmental problems and the way in which that action cuts though the political touch on precisely what the symbolic fantasy order of everyday politics disavows. However, this action only appears as the Real of the dominant symbolic order, for the Real is also that which changes the coordinates or fundamental nature of the order, "changing the entire field of what is acceptable" and what exists. See Slavoj Žižek, "Class Struggle or Postmodernism? Yes, Please!," in *Contingency, Hegemony, Universality* (London: Verso, 2000), 121–25.

41. Ibid.

42. Slavoj Žižek, "Class Struggle or Postmodernism?," in *Contingency, Hegemony, Universality: Contemporary Dialogues on the Left*, ed. Slavoj Žižek, Ernesto Laclau, and Judith Butler (London: Verso, 1999), 121.

43. Jacques Derrida, "Ethics and Politics Today," in *Negotiations: Interventions and Interviews, 1971–2001* (Stanford, Calif.: Stanford University Press, 2002), 298.

44. Ibid.

45. Ibid.

46. Online Etymology Dictionary, http://www.etymonline.com.

47. The Center for Naval Analysis Corporation, *National Security and the Threat of Climate Change*, 13.

48. Air University, *Air University Space Primer* (Montgomery, Ala.: 2003), 3–1, http://www.au.af.mil/au/awc/space/primer. See especially chap. 3, "Space Operations and Tactical Application—U.S. Navy," http://www.au.af.mil/au/awc/space/primer/navy_space.pdf. See also the Federation of American Scientists website for more info: http://www.fas.org. The command appears to have become the Space and Naval Warfare Systems Command: "Navy Intelligence and Security Doctrine," http://fas.org/irp/doddir/navy.

49. The Center for Naval Analysis Corporation, *National Security and the Threat of Climate Change*, 14.

50. Ibid., 39–40.

51. Ibid., 41.

52. Walden Bello, "Conclusion: From American Lake to a People's Pacific in the Twenty-First Century," in *Militarized Currents: Toward a Decolonized Future*

in Asia and the Pacific, ed. Setsu Shigematsu and Keith L. Camacho (Minneapolis: University of Minnesota Press, 2010), 309.

53. Ibid., 308.

54. Ibid., 311.

55. "U.S. to Resume Military Aid for Indonesia," AFX News Limited, November 22, 2005, http://www.forbes.com.

56. Bello, "Conclusion," 312.

57. Parsons, "Taking Up the Security Challenge of Climate Change" (Carlisle, Pa.: Strategic Studies Institute, August 2009), 5.

58. The report was apparently originally available to the public, but the web address given in Parson's report is no longer available. Research on the CNA website will produce the title of the report, but the page states that the report is not for public viewing. See http://www.cna.org.

59. Parsons, "Taking Up the Security Challenge of Climate Change," 16.

60. Ibid.

61. Michael Hardt and Antonio Negri, *Multitude: War and Democracy in the Age of Empire* (New York: Penguin, 2004), xiii.

62. Ibid., 3.

63. Ibid., 12.

64. Ibid., 15.

65. See Richard Howard, trans., *Madness and Civilization* (New York: Random House, 1965).

66. For more information, see the NSCE conference website for the 2012 conference: http://www.environmentandsecurity.org. Information concerning Admiral Titley's remarks comes from my personal attendance.

67. Paul Virilio, *The Original Accident*, trans. Julie Rose (Cambridge, U.K.: Polity, 2007), 11.

Conclusion

1. Alexandre Kojève, *Introduction to the Reading of Hegel*, ed. Allan Bloom, trans. James H. Nichols Jr. (Ithaca, N.Y.: Cornell University Press, 1980), 159. Quoted in Giorgio Agamben, *The Open: Between Man and Animal* (Stanford, Calif.: Stanford University Press, 2004), 9.

2. Francis Fukuyama, *The End of History and the Last Man* (New York: Simon and Schuster, 1992).

3. Agamben, *The Open*, 10.

4. Ibid., 10.

5. Kojève, *Introduction to the Reading of Hegel*, 161. Quoted in Agamben, *The Open*, 10.

6. Agamben, *The Open*, 13.

7. There is evidence for such a claim in Agamben's more recent work. See Giorgio Agamben, *The Kingdom and the Glory: For a Theological Genealogy of Economy and Government (Homo Sacer II, 2)* (Stanford, Calif.: Stanford University Press, 2011). A posthistorical ideology such as Thatcher's form of neoliberalism, it should be remembered, did not dismiss religion. It found in a certain constitution of religion, still powerful in late twentieth-century observances, an absolute morality that matched the absolute character of her authoritarian populism. Thatcherism located in the absolutist strains of religion a theological basis for redeeming the logic of late twentieth-century capitalism. Christianity, like capitalism's promotion of rugged individualism and hard-working entrepreneurialism (now understood as the basic struggle for life), was "about spiritual redemption, not social reform." Life, for Thatcher, was about the combination of "morality" with "work" (work life), and this combination is made possible in and through the realization of "abundance"—in other words, the "high yield" dictated by the discourse of enclosure: "I believe that by taking together these key elements from the Old and New Testaments, we gain: a view of the universe, a proper attitude to work, and principles to shape economic and social life. We are told we must work and use our talents to create wealth. 'If a man will not work he shall not eat' wrote St. Paul to the Thessalonians. Indeed, abundance rather than poverty has a legitimacy which derives from the very nature of Creation." "Creation" itself—the idea of Nature as originating and controlled by God—is resituated within the capitalist drive for abundance and accumulation: the very neoliberal ideological conceptions of the basic life struggle. Margaret Thatcher, "Speech to the General Assembly of the Church of Scotland," May 21, 1988, Margaret Thatcher Foundation, http://www.margaretthatcher.org.

8. Ibid., 40.

9. This is actually an observation of Uexküll's highlighted by Deleuze and Guattari.

10. Félix Guattari and Gilles Deleuze, *A Thousand Plateaus*, trans. Brian Massumi (Minneapolis: University of Minnesota Press, 1987), 257.

11. Salman Rushdie, *The Ground beneath Her Feet* (New York: Picador, 1999), 506.

12. Agamben, *The Open*, 41.

13. Ibid.

14. Ibid., 51.

15. Ibid.

16. It should be noted here that Heidegger's specific formulation of this idea of the animal arises from an experiment with an *insect*: the famous example of the bee, which he takes from an experiment described by Uexküll. In the experiment, the bee is presented with an overflowing amount of honey. After it begins to feed, its abdomen is cut away. But the bee continues to feed without changing its behavior. Hence

the example of the "animal" here is highly problematic, given that it is presented in a *categorical* fashion rather than according to the nomadic and heterogeneous manner of the very concept of the *Umwelt* that Heidegger is attempting to construct and exemplify. Nonetheless, the point is to consider the division being made between Man and Animal and how that division is being challenged by Heidegger in terms other than linguistic and other than the classical tradition of *animal rationale*.

17. Martin Heidegger, *The Fundamental Concepts of Metaphysics: World, Finitude, Solitude*, trans. William McNeill and Nicholas Walker (Bloomington: Indiana University Press, 1995), 263. Quoted in Agamben, *The Open*, 52.

18. Martin Heidegger, "The Origin of the Work of Art," in *Poetry, Language, Thought*, trans. Albert Hofstadter (New York: Harper and Row, 1971), 46. Quoted in Agamben, *The Open*, 71.

19. Ibid., 47.

20. Agamben, *The Open*, 77.

21. Ibid., 80.

22. Ibid.

23. J. V. Beckett, "Landownership and Estate Management," in *Agrarian History of England and Wales*, vol. 4 (Cambridge: Cambridge University Press, 1989), 547.

24. "Press Briefing by Special Rapporteur on Right to Adequate Housing," http://www.un.org.

25. David A. Victor, "Brazil, Doing Business In," in *Encyclopedia of Business*, 2nd ed., http://www.referenceforbusiness.com.

26. Ibid. See also "Habitat Press Release: One Hundred Million Are Homeless in World," http://www.un.org. See as well "Mumbai Takeover: Record 6,000 Shanties Flattened in a Day, Next in Line, Illegal 'Well-Off,'" http://www.indianexpress.com.

27. Jenniver Franco and Saturnino M. Borras Jr., eds., *Land Concentration, Land Grabbing and People's Struggles in Europe* (Amsterdam, The Netherlands: Transnational Institute, June 2013), http://www.tni.org.

28. UNHCR: The UN Refugee Agency, "Climate Change Could Become the Biggest Driver of Displacement: UNHCR Chief," December 16, 2009, http://www.unhcr.org.

29. International Organization for Migration, "Migration, Climate Change and the Environment," n.d., http://www.iom.int.

30. Environmental Justice Foundation, "Climate Justice: Protecting Climate Refugees," London, U.K., http://ejfoundation.org/campaigns/climate/item/climate-justice-protecting-climate-refugees.

31. Ibid.

32. Alain Badiou, *Ethics: An Essay on the Understanding of Evil* (New York: Verso, 2001), esp. 41–44, 70.

33. For the full discussion of "the Name of Thought," see Bill Readings, *The University in Ruins* (Cambridge, Mass.: Harvard University Press, 1996), 159–65.

34. E. Chivian, and A. Bernstein, eds., *Sustaining Life: How Human Health Depends on Biodiversity* (New York: Oxford University Press, 2008), 22, 39. See also C. D. Thomas, A. Cameron, R. E. Green, M. Bakkenes, L. J. Beaumont, Y. C. Collingham, B. F. N. Erasmus, M. Ferreira de Siqueira, A. Grainger, L. Hannah, L. Hughes, B. Huntley, A. S. van Jaarsveld, G. F. Midgley, L. Miles, M. A. Ortega-Huerta, A. T. Peterson, O. L. Phillips, and S. E. Williams, "Extinction Risk from Climate Change," *Nature* 427 (2004): 145–48.

35. Susan Solomon, Gian-Kasper Plattner, Reto Knutti, and Pierre Friedling-stein, "Irreversible Climate Change Due to Carbon Dioxide Emissions," *PNAS* (*Proceedings of the National Academy of Sciences of the United States of America*) 106, no. 6 (2009): 1704–9.

36. http://co2now.org.

37. Ibid.

38. Environmental Protection Agency, "Future Climate Change," http://www.epa.gov.

39. Zeke Hausfather, "Global CO_2 Emissions: Increases Dwarf Recent U.S. Reductions," Yale Forum on Climate Change and the Media, 2013, http://www.yaleclimatemediaforum.org.

40. Jos G. J. Olivier, Greet Janssens-Maenhout, and Jeroen A. H. W. Peters, *Trends in Global CO_2 Emissions: 2012 Report* (The Hague/Bilthoven: Netherlands Environmental Assessment Agency, 2012), http://edgar.jrc.ec.europa.eu/CO2REPORT2012.pdf.

41. Steven L. Simon, André Bouville, and Charles E. Land, "Fallout from Nuclear Tests and Cancer Risks," *American Scientist* 94, no. 1 (2006): 48–57, 48.

42. Comprehensive Nuclear-Test-Ban Treaty Organization, "General Overview of the Effects of Nuclear Testing," 2012, http://www.ctbto.org.

43. Arjun Makhijani, Howard Hu, and Katherine Yih, *Nuclear Wastelands: A Global Guide to Nuclear Weapons Production and Its Health and Environmental Effects* (Boston: MIT Press, 2000), 587.

44. Barry S. Levy and Victor W. Sidel, *War and Public Health* (Oxford: Oxford University Press, 2008), 160.

45. International Physicians for the Prevention of Nuclear War, *Radioactive Heaven and Earth: The Health and Environmental Effects of Nuclear Weapons Testing in, on, and above the Earth* (New York: Apex, 1991), 40.

46. Wuyi Omitoogun and Elisabeth Sköns, "Military Expenditure Data: A 40-Year Overview," in *SIPRI Yearbook 2006: Armaments, Disarmament and International Security* (Oxford: Oxford University Press, 2006), 269.

47. Ibid., 275.

48. Dan Smith, *State of the World Atlas* (New York: Penguin, 1998), 54; Shekhar Gupta, *The Military Balance 1998/99* (Oxford: Oxford University Press, 1998), 14.

49. John Mueller and Mark G. Stewart, *Terror, Security, and Money: Balancing the Risks, Benefits, and Costs of Homeland Security* (New York: Oxford University Press, 2011), 1.

50. Elisabeth Sköns, Wuyi Omitoogun, Catalina Perdomo, and Petter Stålenheim, "Military Expenditure," in *SIPRI Yearbook 2005: Armaments, Disarmament and International Security* (Oxford: Oxford University Press, 2006), 307.

51. Allison J. K. Bailes, "Global Security Governance: A World of Change and Challenge," in *SIPRI Yearbook 2005: Armaments, Disarmament and International Security* (Oxford: Oxford University Press, 2006), 4.

52. Omitoogun and Sköns, *SIPRI Yearbook 2005*, 287.

53. The White House, *The National Security Strategy of the United States of America* (Washington, D.C.: September 2002), 15, http://www.whitehouse.gov.

54. Omitoogun and Sköns, *SIPRI Yearbook 2005*, 287.

55. *U.S. Department of State, Proliferation Security Initiative Participants* (Washington, D.C.: Bureau of International Security and Nonproliferation, 2014), http://www.state.gov/t/isn/c27732.htm.

56. Ibid., 25.

57. Ibid., 21.

58. Ibid., 491.

59. John W. Dower, *Cultures of War: Pearl Harbor: Hiroshima: 9–11: Iraq* (New York: W. W. Norton, 2011), 222.

60. Ibid., 222–23.

61. Martin Heidegger, "The Age of the World Picture," in *The Question Concerning Technology and Other Essays* (New York: Harper and Row, 1977), 135.

62. Edward W. Said, *Culture and Imperialism* (New York: Alfred A. Knopf, 1993), 332.

63. Donald E. Pease, "The Global Homeland State: Bush's Biopolitical Settlement," *boundary 2* 20, no. 3 (2003): 1–18, 7.

64. Ibid., 22.

Index

ABC News, 1, 74, 271n
Accidental, the, 31–40, 42–56, 58,
   60–69, 71–72, 74–78, 96, 131, 136,
   142–43, 150, 180, 182, 184, 194–95,
   197, 208–14, 216, 218–19, 221,
   224–28, 250, 252, 255, 257, 259–60.
   See also Virilio, Paul
aerial photography, 45, 47, 189, 192
Agamben, Giorgio, 27, 56, 61, 64,
   69, 90, 105, 196, 199, 204–5, 213,
   224, 229, 232–37, 241–42, 244–45,
   259–60, 265n, 269n, 275n, 289n,
   291n, 293n, 294n, 295n; kenomatic
   state, 204, 205, 215, 229, 237
Althusser, Louis, 36, 62, 182, 269n, 270n
American exceptionalism, 80, 196,
   198–99
Anderson, Benedict, 55, 121
Anderson, Kym, 107, 280n
anthropocene, 27, 29, 147, 149, 158,
   231–32, 241–42, 245
army field manual, 8
Arnold, Matthew, 51
artificial intelligence, 165, 193
atomic/nuclear arms, 11, 21, 31, 34,
   35–36, 40–42, 54–55, 60, 80–81,
   147, 152–54, 159, 180, 183, 190, 192,
   209, 229, 252, 254–59

Bacon, Francis, 80
Badiou, Alain, 248, 295n

Bailes, Allison, 257, 297n
Batho, Gordon, 85, 273n
Baudrillard, Jean, 161
Bello, Walden, 222, 223, 292n,
   293n
Bentham, Jeremy, 87, 89
Bercovitch, Sacvan, 80, 196, 199–200,
   272n, 289n
biofuels, 79, 102, 104, 111–12
Blick, Caroline, 105
Borlaug, Norman, 38
Bouville, André, 254–55, 296n
Brazil, 19, 62, 67, 72, 76, 78, 82, 83, 87,
   96–113, 124, 128, 130–35, 140, 143,
   246, 249
Brazilian Landless Workers Move-
   ment (MST), 83, 96–98, 101, 107–8,
   124–25, 128–30, 134–44, 217, 246,
   249, 267, 269, 278, 282, 283, 295
Bush, George W.: Bush administration,
   3, 6–7, 27–29, 56, 183, 195–99, 222,
   228, 256
Bush, Vannevar, 52, 180–82, 287n
Butler, Judith, 45, 48, 266n, 292n

Capra, Fritjof, 46, 206
Carson, Rachel, 2, 21, 263
Carter, Jimmy: Carter administration,
   185, 255
Center for a New American Security
   (CNAS), 1, 5, 153, 209, 271n

ROBERT P. MARZEC is associate professor of ecocriticism and postcolonialism in the Department of English at Purdue University. He is the author of *An Ecological and Postcolonial Study of Literature*, editor of *Postcolonial Literary Studies: The First Thirty Years*, and associate editor of the journal *Modern Fiction Studies*.